ANTHONY ROBINSON
AIRCRAFT OF WORLD WAR 3

ANTHONY ROBINSON

AIRCRAFT OF WORLD WAR 3

CRESCENT
New York

A Bison Book

This edition is distributed by
Crescent Books,
a division of Crown Publishers, Inc.
One Park Avenue,
New York New York 10016

Produced by Bison Books Corp
17 Sherwood Place
Greenwich, CT 06830. USA.

Library of Congress Cataloging in Publication Data

Robinson, Anthony,
   Aircraft of World War III.
   1. Airplanes, Military.   2. North Atlantic Treaty
Organization.   3. Warsaw Treaty Organization.   4. World
War III.   I. Title.   II. Title: Aircraft of World
War 3.   III. Title: Aircraft of World War Three.
UG1240.R6   1983   358.4'183   82-23573
ISBN 0-517-37999-6

Printed in Hong Kong

Half title: the Tornado air defense variant carries four Skyflash air-to-air missiles beneath the fuselage.
Title: US Navy A-7E attack aircraft fly in formation.
Contents: The AV-8B Harrier II will equip all US Marine Corps light attack squadrons.

List of Abbreviations

AAM Air-to-Air Missile
ABM Anti-Ballistic Missile
Ace (Mobile Force) Allied Command Europe
ACMI Air Combat Maneuvering Instrumentation
ADCOM Aerospace Defense Command
ADGE Air Defense Ground Environment
ADIZ Air Defense Identification Zone
AEW Airborne Early Warning
AFB Air Force Base
AFRES Air Force Reserve (US)
AFSATCOM Air Force Satellite Communications (system)
ALAT Aviation Légère de l'Armée de Terre (French army light aviation)
ALCM Air-Launched Cruise Missile
AMRAAM Advanced Medium Range AAM
ANG Air National Guard
ARM Anti-Radiation Missile
ARRS Aerospace Rescue and Recovery Vehicle
Asat Anti-satellite
ASM Air-to-Surface Missile
ASPJ Advanced Self-Protection Jamming (system)
ASW Anti-Submarine Warfare
AWACS Airborne Warning and Control System
AWADS Adverse Weather Aerial Delivery Device
BMD Ballistic Missile Defense
BMEWS Ballistic Missile Early Warning System
BM Bombardment Wing
C³ Command Control and Communications
CAP Combat Air Patrol
CAS Close Air Support
CBU Cluster Bomb Unit
CEP Circular Error Probable
CSIRS Covert Survivable-in-Weather Reconnaissance/Strike (fighter)
DEW Distant Early Warning
DSCS Defense Satellite Communications System
ECM Electronic Counter Measures (also ECCM Counter-Countermeasures)
ELINT Electronic Intelligence
EMP Electro-Magnetic Pulse
ESM Electronic Support Measures
ESSS External Stores Support System
EVS Electro-optical Viewing System
EW Electronic Warfare
FAC Forward Air Control
FACTS Fuel And Sensor Tactical (packs)
FFAR Folding Fin Aircraft Rocket
FLIR Forward Looking Infra-Red
FOBS Fractional Orbit Bombardment System
GLCM Ground Launched Cruise Missile
GPES Ground Proximity Extraction System
GPS Global Positioning System
HARM High-speed ARM
hi-lo-hi Describing the flight path of an aircraft on a war mission. In this case the aircraft flies to the target area at high altitude, makes the attack at low level, and returns to base at high altitude. Variations eg lo-lo-lo are possible
HUD Head Up Display
ICBM Intercontinental Ballistic Missile
JATO Jet-Assisted Take Off
JTIDS Joint Tactical Information Distribution System
LANTIRN Low Altitude Navigation and Targeting Infrared System for Night
LAPES Low Altitude Parachute Extraction System
LLTV Low Light Level Television

Contents

1: Strategic Forces   6
2: War in Space   42
3: Tactical Combat   54
4: War At Sea   92
5: Air Defense   114
6: Recce and ELINT   130
7: Transport Aircraft   148
8: Army Aviation   166
9: The Balance of Forces   182
Index   191

LORAN Long Range Navigation (equipment)
LRA Long Range Aviation (Soviet)
MAD Magnetic Anomaly Detector
MIRV Multiple Independently-Targeted Re-entry Vehicle (also MRV ie not independently targeted)
MPS Multiple Protective Structures
NATO North Atlantic Treaty Organization
NAVWASS Navigation and Weapons' Aiming Subsystem
NBC Nuclear Bacteriological or Chemical
ORI Operational Readiness Inspection
PACAF Pacific Air Forces
PGM Precision Guided Munitions
PNVS Pilot's Night Vision Sensor
RPV Remotely Piloted Vehicle
SAC Strategic Air Command
SAM Surface-to-Air Missile
SAR Search and Rescue
SATS Short Airfield for Tactical Support
SDS Satellite Data System
SLAR Sideways-looking Airborne Radar

SLBM Submarine Launched Ballistic Missile
SMAC Scene-Matching Area Correlator
STAR Surface-to-Air Recovery (system)
STOL see V/STOL
SRAM Short Range Attack Missile
TAC Tactical Air Command
TACAMO Take Command and Move Out (aircraft)
TACAN Tactical Air Navigation (system)
TADS Target Acquisition Designation Sight
TARPS Tactical Airborne Reconnaissance Pod Sytem
TERCOM Terrain Contour Matching
TEREC Tactical Electronic Reconnaissance
TFW Tactical Fighter Wing
TISEO Target Identification System Electro-Optical
TOW Tube-Launched, Optically-tracked, Wire-guided (antitank missile)
TRAM Target Recognition Attack Multisensor
VAS Visual Augmentation System
VLF Very Low Frequency
V/STOL Vertical or Short Takeoff and Landing

The USAF's new intercontinental-range bomber, the Rockwell
B-1B, is to become fully operational with Strategic Air Command
in 1988. A B-1A prototype is pictured during a test flight in 1975.

6

1. STRATEGIC FORCES

Strategic Forces

Delivery systems for strategic nuclear weapons fall into three main categories: land-based missiles, submarine-launched missiles and bomber aircraft (which can carry air-launched missiles). The manned bomber concept is of course the oldest of the nuclear delivery systems. It cannot compete with the land-based missile in speed of delivery combined with accuracy. Nor at present does it compare with the submarine-launched ballistic missile, which has a near invulnerability, because of the difficulties of detecting the launching craft. It is therefore tempting to dismiss the bomber as an anachronism, which owes its survival to military conservatism and the tendency of defense planners to over-insure.

The United States is nonetheless committed to maintaining the 'triad' of land-based ICBMs (intercontinental ballistic missiles), SLBMs (submarine-launched ballistic missiles) and strategic bombers. The Soviet Union likewise divides its nuclear delivery systems between ICBMs, SLBMs and bombers and both the Superpowers are seeking to modernize their strategic bomber forces.

One of the reasons for retaining the bomber force is the fear that silo-based ICBMs may become vulnerable to attack from enemy ICBMs. Similar concern is voiced about possible developments in ballistic missile defense technology, which could neutralize both land and submarine-based missile systems. It is therefore prudent to maintain a strategic nuclear delivery system which will be unaffected by such developments.

Bombers have other positive virtues when compared with missile systems. They can be dispersed from their bases, or put on airborne alert, thus warning the enemy, in time of crisis, of one's resolution to act and also denying him the option of a pre-emptive strike against the bomber force. A single bomber can carry a heavy load of free-fall or stand-off weapons, enabling it to attack a variety of targets in the same mission. It can be diverted after takeoff to a new target of greater importance than its original objective and it can be recalled if the order to attack proves to be a miscalculation. The bomber's crew can avoid wasting warheads on a target already devastated and they are able to make some assessment of the effectiveness of their attack. All of this represents a flexibility which is unobtainable from an all-missile force. Finally bombers oblige the enemy to devote valuable resources to his air defenses.

In the United States both the ICBM force and the bomber wings are controlled by the USAF's Strategic Air Command (SAC). With its headquarters at Offutt Air Force Base (AFB) Nebraska, this command operates from more than 50 bases and controls approximately 120,000 personnel. The ballistic-missile-armed submarines (SSBNs) are the responsibility of the US Navy. The Navy's aircraft carriers also have a measure of strategic capability by virtue of their nuclear-capable attack aircraft, although these are primarily tactical weapons.

The mainstay of SAC's bomber force at present is the Boeing B-52 Stratofortress. This giant eight-engined bomber first entered service in 1955, but only the last two production variants (the B-52G and B-52H) now serve in the strategic role. A third version, the B-52D, has been retained because of its enhanced conventional bomb load, but, although it has a nuclear capability, its importance is primarily tactical.

A total of 295 B-52Gs and B-52Hs was built (193 Gs and 102 Hs), with the last delivered in June 1962, and 269 remain in service with 14 SAC bombardment wings. The B-52G has a span of 185 feet, a length of 160ft 11in and maximum all-up weight is 488,000 pounds. It is powered by eight 13,750lbs thrust Pratt & Whitney J57-P-43WB turbojets. Fuel capacity is almost 48,000 gallons, which gives the B-52G an unrefueled range of 8400 miles. The B-52H, with the same fuel capacity, has an unrefueled range of 10,130 miles, because of the greater efficiency of its Pratt & Whitney TF33 turbofan engines. Its maximum speed of 545 knots is some five knots below that of the B-52G.

The normal crew complement of the Stratofortress is six, comprising pilot, co-pilot, navigator, radar navigator, electronic warfare officer and tail gunner. The radar navigator directs the bomb run and prepares the weapons for release. The tail gunner is located in the forward crew compartment and operates his turret by remote control. The B-52G has four 0.5in machine guns in the tail, whereas the B-52H mounts a 20mm Vulcan rotary cannon, capable of firing 4000 rounds per minute. The gun turret may seem out of place in the age of the air-to-air missile, but many Soviet warplanes are so armed, and the B-52's tail guns accounted for two MiG-21 interceptors during the Vietnam War.

The B-52Gs and Hs can carry a warload of up to eight nuclear free-fall bombs. They can also be armed with up to 20 SRAM (short-range attack missile) rounds, eight carried internally in a revolving cylinder in the bomb bay, with the remainder on wing pylons. The SRAM has a nuclear warhead of some 200 kilotons (kt) yield and has a range of between 35 and 105 miles, depending on the altitude of the launch aircraft and the preprogrammed maneuvers of the missile. In practice B-52s are likely to be armed with a combination of free-fall weapons and SRAMs. The latest addition to the Stratofortress' armory is the Boeing AGM-86B air-launched cruise missile (ALCM), with a range of 1500 miles and a 200 kt warhead. Current plans call for all 173 remaining B-52Gs to be modified to carry up to 20 AGM-86Bs each (eight internally and twelve on wing pylons). Early conversions will carry only the wing-mounted missiles and the first squadron is scheduled to become operational at the end of 1982. The B-52Gs may be followed by 96 B-52H ACLM carriers.

As originally conceived, the B-52 was to rely on its high-altitude performance to penetrate enemy defenses, as the name Stratofortress implies. Yet by 1959 it was recognized that developments in surface-to-air

A Boeing B-52G lifts off from Seymour-Johnson AFB, NC, home of the 68th Bomb Wing, during SAC's world-wide readiness exercise Global Shield in July 1979.

missile (SAM) technology (later paralleled by improvements in manned interceptors) had made such tactics unworkable. The problem was compounded by the massive bomber's conspicuous radar 'signature'.

The solution was to switch to low-level penetration at high subsonic speed. However, the B-52 was ill-suited to such tactics. Its large, high-aspect-ratio wing is ideal for the still air conditions of high-altitude flight, but badly adapted to cope with the turbulence met at low level. This results in excessive loads on the airframe and control difficulties. Consequently a series of expensive modifications has been carried out to improve the aircraft's structural integrity and to prolong airframe life. In addition a stability augmentation system has been fitted to improve controllability in turbulence and to ease structural loads. Other essential modifications for the low-level mission include the fitting of terrain avoidance radar and an improved radar altimeter. Suitably modified B-52Gs and B-52Hs can now operate down to an altitude of 300ft, earlier models being restricted to 500ft.

Flying at low level is not by itself enough to ensure the B-52's survival in hostile airspace. Electronic countermeasures (ECM) have an important part to play in neutralizing enemy defenses and are continually being upgraded to meet new threats. The latest ECM

update (Phase VI) for the B-52G and H was begun in 1974 at a cost of $362,500,000. A separate modification program warns of enemy attack through the ALQ-153 tail warning radar, which can detect SAMs, interceptors and air-to-air missiles (AAMs).

Forward visibility at low level was found to be a problem and to supplement the terrain avoidance radar the B-52Gs and Hs have been fitted with an electro-optical viewing system (EVS). This comprises a steerable low-light television (LLTV) and a forward-looking infra-red (FLIR) mounted in blisters under the bomber's nose. The view from either sensor can be shown on displays on the flight deck, warning of such hazards as radio masts. The EVS is also useful for picking up navigational waypoints, or targets and for making a post-attack assessment.

The difficulty of accurate navigation and target acquisition at low level is being tackled by a massive two-stage Offensive Avionics Systems program. This will give the B-52Gs and Hs a greatly enhanced capability through improvements in radar, navigation aids and computers of at least 30 percent. The B-52's AQS-38 forward-looking radar is modified by adding a digital processor in the first stage. Stage two may involve the substitution of an entirely new radar, which could incorporate the latest advances, including frequency agility to counter enemy jamming. It could also give the bomber terrain-following capability (the present system only provides for terrain avoidance), but this would involve a major redesign of the B-52's flight control

system. Whether the remaining useful life of the B-52 would justify such expense is doubtful, particularly in view of the decision to introduce the B-1B and a 'Stealth' bomber into SAC by the 1990s.

The basic aircraft operating unit within SAC is the bombardment wing (BW). B-52 wings comprise one or sometimes two squadrons flying the bombers and one or two air refueling squadrons operating Boeing KC-135A tanker aircraft. The strategic bombers rely on air refueling to execute their mission and the importance of this support is emphasized by the inclusion of air refueling squadrons within each bombardment wing. The B-52G/H force was assigned to the following wings in early 1982:

Unit	Base	Composition
2nd BW	Barksdale AFB, La	Two sqns of B-52Gs Two sqns of KC-135As
5th BW	Minot AFB, ND	One sqn of B-52Hs One sqn of KC-135As
19th BW	Robins AFB, Ga	One sqn of B-52Gs One sqn of KC-135As
28th BW	Ellsworth AFB, SD	Two sqns of B-52Hs One sqn of KC-135As One sqn of EC-135s (Flying command posts)
42nd BW	Loring AFB, Maine	One sqn of B-52Gs Two sqns of KC-135As
68th BW	Seymour Johnson AFB, NC	One sqn of B-52Gs One sqn of KC-135As
92nd BW	Fairchild AFB, Wa	One sqn of B-52Gs Two sqns of KC-135As
93rd BW	Castle AFB, Ca	Two sqns of B-52Gs & Hs Two sqns of KC-135As
97th BW	Blytheville AFB, Ar	One sqn of B-52Gs One sqn of KC-135As
319th BW	Grand Forks AFB, ND	One sqn of B-52Hs One sqn of KC-135As
320th BW	Mather AFB, Ca	One sqn of B-52Gs One sqn of KC-135As
379th BW	Wurtsmith AFB, Mi	One sqn of B-52Hs One sqn of KC-135As
410th BW	KI Sawyer AFB, Mi	One sqn of B-52Hs One sqn of KC-135As
416th BW	Griffiss AFB, NY	One sqn of B-52Gs One sqn of KC-135As

A proportion of the B-52 force is always on ground alert. This would typically involve six bombers and three tankers in a wing, ready to take off at a moment's notice. The B-52s are parked on hardstandings positioned off a central taxiway, which leads onto the main runway. They are armed and fueled, have been inspected by their crews and prepared for a quick engine start. Each is fitted with individual cartridge starters on all eight engines to speed up this process.

Once the klaxon sounds the crew are driven to their aircraft, while the groundcrew remove engine inlet covers. The bombers take off first, followed by the tankers. The alert force will be followed by as many of the remaining aircraft in the wing as can be readied for combat. SAC crews are trained in minimum interval takeoffs, which allow only some 15 seconds between each aircraft. This enables the alert force to be airborne within three minutes of engine start, essential if the air base is threatened by missile attack. The B-52's unique landing gear enables it to cope with winds blowing across the runway by allowing the aircraft to slew into the wind, while the undercarriage stays lined-up with the runway centerline.

Before ground alert procedures were refined, it was deemed necessary to keep a proportion of the bomber force on a continuous airborne alert. This involved a force of bombers flying holding patterns at high level over Alaska, Canada and the North Atlantic, within striking distance of their assigned targets. This procedure was expensive in fuel and aircraft maintenance. It was also tiring for crews, who were in the air for up to 24 hours and had to fly precisely to avoid wasting fuel. Despite its disadvantages, airborne alert is useful at time of crisis and SAC retains the capability to reintroduce it at a moment's notice.

When the SAC bomber force has been launched on its mission, the aircraft fly to a predetermined point on their track to the target. There they must fly a holding pattern until they receive further orders to proceed or to return to base. Coded orders to execute the attack can only come from the National Command Authorities, the President or those authorized to act on his behalf. Thus positive control of the bomber force is ensured and the procedure can be regarded as 'fail-safe', because if no further orders are received the bombers will abandon the mission. Communications are therefore of vital importance to SAC and since 1979 the B-52 force has been fitted with AFSATCOM, the USAF's satellite communications system, which provides high-priority communications anywhere in the world.

The performance of SAC's air and maintenance crews and the effectiveness of the operating procedures are constantly monitored and evaluated by an exhaustive series of inspections. Practice alerts are held to test reaction times and may be terminated once the engines have been started or the first bomber has

A B-52's awesome bomb load explodes on a Viet Cong base camp in South Vietnam (above left). The Guam and Thailand-based B-52Ds (below) could lift 108 750lb bombs in a single mission. A B-52G crew (left) board their aircraft during a practice alert in the US.

reached the runway. An aircraft's first sortie after ground alert is an important check on the standard of maintenance during its time on the ground. The crew is also assessed during training flights, which include simulated bombing attacks.

On a larger scale, the wing may be subjected to a no-notice Operational Readiness Inspection (ORI) by a highly-critical team from SAC Headquarters. This would involve every available aircraft flying a simulated war mission. In March 1980 the crews of two B-52Hs from the 410th BW were ordered to carry out a surveillance mission over the Persian Gulf, with no prior notice. This mission required a 43½-hour non-stop flight, with five inflight refuelings, a convincing demonstration of the B-52's global reach.

The demanding low-level bombing mission is practiced over specially selected routes in the USA. The B-52s are flown at heights of between 1000 and 2000ft, until the defense penetration phase of the exercise, lasting for perhaps an hour, when the bombers descend to between 500 and 300ft. The routes are varied to give the crews experience over unfamiliar terrain and navigation is often an exacting task, with little information available from radar returns. Simulated attacks are made on radar bomb-scoring sites, which are maintained by the 1st Combat Evaluation Group. This unit also monitors the accuracy of the bombers' and tankers' navigation throughout the exercise. Bomb release is

simulated by the attacking aircraft transmitting a 'bomb tone', which is cut off at the point of release. The radar bomb scoring site can then compute the point of impact. Several such targets are often attacked on one training mission.

Not all of SAC's training takes place in the air, however. In common with other air forces, the USAF has been affected by the energy crisis and this means that the service is looking for substantial fuel economies. Flight simulators are one answer and have the added advantages that they conserve airframe hours and enable emergency procedures to be practiced in complete safety. The latest B-52 simulator offers the flight deck crew six-axis motion, computer generated image visual displays and simulated control force feel. The radar navigator can receive simulated radar returns and can practice weapons release procedures. Numerous simulated 'threats' can exercise the electronic warfare officer in the use of ECM equipment. Yet despite this sophistication, simulators do not offer an entirely satisfactory substitute.

Another way of saving fuel is to carry out a part of the training program in a less 'thirsty' aircraft than the B-52. It has been proposed that SAC buy about 60 business-jet aircraft, in which the bomber crews (less the gunner) could fly 25 percent of the training sorties

SAC manages all USAF tanker aircraft, including this KC-135A (left) fitted with a drogue adaptor. The FB-111 (below) has just launched a SRAM missile.

now undertaken by the B-52. The annual fuel savings of this measure are estimated as 100,000,000 US gallons. At present B-52 flights for co-pilot proficiency training are being reduced by using T-37 and T-38 trainers for this purpose.

Since 1948 SAC has held an annual bombing competition in which every wing takes part. As with much of SAC's peacetime activity, the object of the competition, code-named 'Giant Voice', is to test efficiency and readiness of bomber units and the tanker force. The B-52-equipped wings are the primary contenders, but SAC's two FB-111 wings also take part, as do tanker squadrons from the regular air force, the Air Force Reserve (AFRES) and Air National Guard (ANG). Among the skills tested are high and low-level bombing accuracy, use of ECM, evasion of fighter interception, navigation, air refueling and SRAM launches.

Exercise 'Global Shield' is a command-wide, no-notice readiness exercise, which involves all SAC units – including missile wings and strategic reconnaissance wings – in a simulated thermonuclear war. 'Global Shield 80' held in June 1980 involved 100,000 SAC personnel and 44 bases in an exercise lasting nine days. It began with an all-out effort to place additional crews and aircraft on alert to participate in the exercise, as of course SAC's actual alert forces could not be used. As would happen in a real period of prewar crisis, a proportion of SAC's bombers and tankers was dispersed to other military airfields or to civil airports. In addition to making the enemy's target planning more difficult, this

provides SAC with more runways and so eases the problem of getting the force off at short notice after an attack warning.

The next stage was an airborne alert and aircraft were required to await positive instructions before carrying out simulated war sorties. This aspect of the exercise differed only in scale from the normal wing ORI. On return from their mission, some aircraft acted as battle damaged bombers which would require special maintenance attention at forward operating locations before they were able to return to base.

As previously noted part of SAC's B-52 force is made up of B-52D aircraft, an older and less capable bomber than the B-52G and B-52H. Although the D model can carry nuclear weapons, its chief usefulness lies in its large conventional bomb load. The normal capacity of the B-52G and H is twenty-seven 750 or 500lb bombs, but the B-52D, by virtue of an internally-modified bomb bay, can lift 108 of these weapons, giving it a nominal bomb load of 70,000lbs. The modification program, named 'Big Belly', was instituted during the Vietnam War and it involved all B-52Ds. Another significant difference is that in the D the gunner is located in the tail turret, some 130 feet behind the flight deck, instead of in the forward crew compartment.

In addition to the 'saturation bombing' role, which was its primary task over Vietnam, the B-52D can undertake sea minelaying and ocean surveillance. Its ECM capabilities are sufficient to allow it to operate over NATO's European fronts. Indeed the bomber has

already faced a severe test in penetrating the North Vietnamese defenses around Hanoi and Haiphong during the 'Linebacker II' strikes in December 1972. These sorties could involve a flying time of 12 hours for the B-52s based on Guam. Fifteen B-52s (nine B-52Ds and six B-52Gs) were lost to an estimated 900 SAMs fired, with AA fire and MiG interceptors scoring no kills. As Linebacker II involved 729 B-52 sorties, 498 of them into the high-risk Hanoi/Haiphong region, the losses are not excessive. They suggest that it would be quite feasible to undertake B-52 sorties over Europe, especially the NATO flank areas, where air defenses would be less intense than on the Central Front and over Soviet territory. The B-52's conventional role is also a useful adjunct to the United States' Rapid Deployment Force.

At the beginning of 1982 there were four wings operating the B-52D in the tactical or conventional strategic role within SAC:

Unit	Base	Composition
7th BW	Carswell AFB, Texas	Two sqns of B-52Ds One sqn of KC-135As
22nd BW	March AFB, Ca	One sqn of B-52Ds One sqn of KC-135As
43rd SW*	Andersen AFB, Guam	One sqn of B-52Ds
96th BW	Dyess AFB, Texas	One sqn of B-52Ds One sqn of KC-135As

*SW = Strategic Wing

An FB-111's weapons bay doors open to reveal a SRAM missile (left). SAC's first variable-geometry wing bomber is dwarfed in comparison with the B-1A which it accompanies in formation (below).

During the course of 1982-83 the B-52D force is to be run down due to the cost of keeping these elderly aircraft in service. However their mission is likely to be taken over by B-52Hs modified to carry up to 108 500lb bombs (84 internally and 24 on wing pylons).

If low-level flying presents problems for the B-52, then by contrast it is the normal operating environment for SAC's other strategic bomber, the General Dynamics FB-111A. This aircraft is a supersonic, variable-geometry-wing, medium range bomber, derived from the F-111 tactical strike fighter. As originally conceived the FB-111 was to have replaced the Convair B-58 Hustler (SAC's first supersonic bomber) and early model B-52s. A total of 263 bombers was planned, but escalating costs and early technical problems with the

F-111 led to a severe curtailment. In the event only 76 FB-111As were built in 1966-71, of which 61 remained operational in 1982.

In spite of its inauspicious beginnings, the FB-111A has proved to be a worthwhile addition to the US strategic armory, largely because of its effectiveness at high speed and low level. The FB-111A-equipped bombardment wings are assigned to targets on the periphery of Soviet territory because the bomber's unrefueled range of 3400 nautical miles classes it as a medium rather than a heavy bomber. Crew fatigue is also a limiting factor, as the FB-111A carries only a pilot and a navigator, compared with the six crewmembers aboard a B-52. However, the bomber is equipped for inflight refueling and tanker squadrons are assigned to the FB-111A wings.

The FB-111A is powered by two Pratt & Whitney TF30-P-7 turbofans with afterburning, each giving 20,350lbs of thrust and maximum speed at altitude is more than Mach 2. The variable-sweep wings endow the FB-111A with excellent handling qualities throughout its speed range. The wings are positioned fully forward, giving a wing span of 70ft, for takeoff, landing and economical cruise. In the fully-swept position, which is selected for high-speed flight, the span is reduced to 33ft 9in. Overall length is 73ft 5in and height is 17ft, thus the FB-111A presents a much smaller target to enemy interceptor pilots and radars than its massive stablemate the B-52. Normal gross weight of the FB-111A is 80,000lbs, which can increase to 110,000lbs maximum gross weight. Weapons carried include the SRAM missile, which can be mounted on the underwing pylons and in the internal weapons bay, which also houses free-fall nuclear weapons. Auxiliary fuel tanks can also be carried on the wing pylons.

The key to the FB-111A's impressive low-altitude performance is the aircraft's complex avionics systems. A terrain-following radar and radar altimeter enable the FB-111A to fly 200ft above the ground, following the contours of the terrain. The system is duplicated and if one should fail the second takes over automatically. In the event of a double failure the aircraft is pulled into a steep climb. Terrain following can be automatic or controlled by the pilot, following cockpit radar displays.

The main navigational aid is an inertial navigation system, which is completely independent of external sources of information and is therefore unjammable. All that is required is that the co-ordinates of the aircraft's starting point are fed into the system's computer. This is so accurate that each aircraft's parking space has been individually surveyed and marked with the exact latitude and longitude. It is claimed that the system could direct a bomber from its hardstanding to the end of the runway even in dense fog. A speed-sensing doppler radar can also be used for navigation and an attack radar is used to acquire the target during the bomb run.

The FB-111A's high speed, low altitude capability provides its best safeguard against enemy defenses. However, the aircraft also carries a range of ECM equipment. This includes a radar homing and warning system (which alerts the crew when their aircraft has been picked up by the enemy), jamming transmitters to interfere with radars and communications, chaff to blot out radar returns and flares to decoy infra-red guided missiles away from the bomber's engine exhausts. Should all these fail, all F-111 variants have a rocket-assisted escape capsule, which will carry the two crew-members to earth in more comfort and safety than the traditional ejection seat and individual parachute.

The two SAC wings which currently operate the FB-111A are:

Unit	Base	Composition
380th BW	Plattsburgh AFB, NY	Three sqns of FB-111As Two sqns of KC-135A/Qs
509th BW	Pease AFB, NH	Two sqns of FB-111As One sqn of KC-135As

There have been various proposals to modernize the existing FB-111A force by re-engining and 'stretching' the surviving airframes to produce a more capable bomber. It has also been suggested that F-111 tactical fighters should be similarly modified to the strategic role to expand the force. However, the USAF has not taken up any of these plans.

One reason for the delay in finding a successor to the B-52 was the change from high to low level target penetration tactics, necessitated by improvements in Soviet air defenses. This led to the cancellation of the North American B-70 in 1964. The B-70 was designed to operate at altitudes of 70-80,000ft at speeds up to Mach 3. It was considered too vulnerable to Soviet missiles and in fact the MiG-25 Foxbat is believed to have been developed specifically to counter the projected B-70 force.

The long-awaited replacement for the B-52 seemed within sight in 1970, when North American Rockwell (later Rockwell International) was awarded a contract to build five test examples of the B-1A bomber to meet the USAF's Advanced Manned Strategic Aircraft requirement. The important characteristic of the new bomber was the ability to fly the target penetration phase of its mission at low altitude and high subsonic speed (Mach 0.85 at 200ft).

A typical mission would involve a scramble takeoff from a base threatened by missile attack, clearing the danger area within five minutes. (The B-52 takes 15 minutes.) In the event of the B-1A being caught in the vicinity of a nuclear explosion, the airframe was 'hardened' to withstand a degree of blast effect and its radar and communications systems were also 'hardened' against the effects of electro-magnetic pulse (EMP). Once airborne the B-1A would rendezvous with a

SAC's long-range bomber force, equipped with elderly B-52s, is to be rejuvenated with the introduction of the B-1 (top). Its Soviet counterpart, Long Range Aviation, at present relies on the 1950s-vintage Tu-20 Bear (right), which is equipped with a tail-mounted gun armament for self-defense. Its turret (above) houses two 23mm cannon.

The Tu-20's stablemate in Long Range Aviation is the turbojet powered M-4 Bison long-range bomber (above). The Tu-16 Badger (above right) has been the mainstay of the Soviet medium-range bomber force since the 1950s. Its intended successor, the Tu-22 Blinder (below) was only produced in limited numbers.

tanker and cruise at high altitude to the edge of the enemy defenses, when it would descend to low altitude for target penetration. The B-1A was intended to carry SRAM missiles for defense suppression and free-fall nuclear bombs to drop on the primary targets. After bomb release it could escape at high speed to a friendly airfield. Its maximum performance was Mach 2.2 at 50,000ft.

Variable-sweep wings gave the B-1A the ability to operate from much shorter runways than the B-52, together with excellent high speed characteristics with the wings fully swept at 67.5 degrees. Power was provided by four GE F101 afterburning turbofans, each giving 30,000lbs thrust at maximum power. The four-man crew comprised pilot, co-pilot and offensive and defensive systems operators. In addition to attack and terrain following radars, the B-1A was equipped with advanced ECM systems.

In 1977 it seemed that the B-1 was to go the same way as the B-70 Valkyrie, when President Carter cancelled the 240-aircraft production program. However flight testing of the four completed prototypes went on, the ECM systems continued to be developed for use on late model B-52s and work continued on the F-101 engine which could also be used to power advanced fighter aircraft. Thus when the Reagan administration came into office with a commitment to rejuvenate American defenses the reinstatement of the B-1 bomber was one option that was available to strengthen the strategic forces. In October 1981 an order for 100 B-1s was announced as part of an ambitious program to modernize the US nuclear arsenal.

The B-1B, which is expected to become fully operational by 1988, differs in several important respects from the earlier B-1A. The gross weight is some 40 tons greater, allowing it to fly a transatlantic mission with full warload and return without inflight refueling. However a more typical mission would be similar to that outlined for the B-1A. Maximum speed is reduced to Mach 1.2, the B-1A's elaborate variable-geometry engine inlets being replaced by simpler fixed inlets. A great advance over the original B-1 model is the reduction of the aircraft's radar signature by a factor of ten, giving it a signature one hundredth that of the B-52.

The B-1B can carry a variety of weapons in three internal bays and on external hard points. Loads of 20-38 free-fall nuclear bombs can be carried, or up to 38 SRAMs, although a mixture of the two is more typical. Looking ahead to the time when the B-1B supplants the B-52 as a cruise missile carrier and conventional bomber, 22 ALCMs can be carried, eight of them in an internal rotary launcher, and up to 128 500lb Mk 82 high explosive bombs can be lifted.

The B-1B's attack and terrain-following radar is based on the F-16 fighter's APG-66 radar. The defensive ECM system, designated ALQ-161, will automatically pick up hostile radar emissions, assign them a priority and initiate jamming or deception against those

19

This cutaway view of the B-1 (left) shows the internal arrangement of the fuselage, with the forward crew compartment, weapons bays and fuel tanks. The United Kingdom retired its Vulcan B Mk 2 bombers (below) in 1982, after a brief period in action during the Falklands conflict. France's Mirage IV strategic bomber (below left) will remain in service as a theater nuclear strike aircraft until the 1990s.

that appear the greatest threat. As the B-1B will be flying fast and low, the ECM system is an additional insurance against enemy interception and missile defenses, which together with the reduced radar signature should ensure its effectiveness well into the 1990s. By this time the new Stealth bomber should be in service and the B-1B will take on the role of cruise-missile carrier and conventional bomber.

By the end of the present decade it is predicted that the Soviet air defenses will begin to introduce advanced new systems capable of detecting, tracking and intercepting low-flying strategic bombers. The key to this enhanced capability is the computer-controlled pulse-doppler radar, able to pick-out low flying aircraft from the ground clutter which hides these from a conventional radar.

The same technology gives a missile-armed interceptor a 'lookdown/shootdown' capability and then the immunity of the low flying bomber disappears. The USAF's answer to this anticipated threat is the Advanced Technology Bomber, which makes use of 'Stealth' technology to hide from enemy radars. These techniques have been applied to the B-1B to reduce its radar signature, but when incorporated from the outset into the new bomber design they are so effective as virtually to eliminate any radar return. The bomber's infra-red signature is also reduced. Indeed it has been claimed that the introduction of the Stealth bomber will make Soviet air defenses obsolete overnight.

The techniques employed to achieve the elimination of an aircraft's radar return are highly classified. Nevertheless a certain amount of general information is available. The starting point of Stealth technology must be the basic design of the airframe. If those parts which act as radar reflectors, such as angular engine intakes and vertical tail surfaces, can be eliminated or reduced in size, then other techniques can suppress the residual reflectivity. It is significant that the contractor for the Advanced Technology Bomber is Northrop, a company with unrivalled experience in 'flying wing' designs, because this configuration comes close to meeting the airframe requirements of a Stealth bomber. If carbonfiber composites replace metals in an aircraft's structure and skinning, they will not reflect radar pulses to the same degree and they also have the incidental advantages of strength and lightness. Finally radar absorbent material can be applied to the remaining surfaces on the aircraft. This will absorb electromagnetic waves and so weaken the return signal.

It is anticipated that the B-1B will find it increasingly difficult to cope with Soviet air defenses in the 1990s. It is therefore likely that the Stealth bomber will

The Soviet Union's Tu-26 Backfire medium-range bomber (left) poses a serious threat to US and allied forces in the European and Far Eastern theaters. Two designs competed to meet the USAF's Air Launched Cruise Missile requirement, the Boeing AGM-86 (upper and lower) beating the AGM-109 Tomahawk (center).

The Soviet strategic forces are divided between three armed services. The air force controls the manned bombers, the navy is responsible for submarine-launched systems and land-based missiles are the province of the Strategic Rocket Forces. This last organization is a service in its own right, quite distinct from the air force, and indeed it ranks above the other four services (Army, Air Defense Forces, Air Force and Navy) in seniority. Thus in contrast to American practice, land-based strategic bombers and missiles fall under two separate commands.

There can be no doubt that, as presently constituted, the Soviet air force's manned bomber arm is the least effective component of the strategic forces. This is due to the lack of a modern bomber of true intercontinental range to replace the 1950s-vintage Tupolev Tu-20 Bear. This type has not had the development potential of the B-52 and is now long overdue for retirement in the demanding role of deep penetration sorties over the continental USA.

Long Range Aviation (LRA) is better equipped for deep penetration missions over such target areas as Western Europe, China, Japan and possibly the periphery of the USA. Its medium range bombers, carrying either nuclear or conventional weapons, could attack such vital military targets as headquarters and supply depots. They could also be targeted against industrial complexes or centers of population. It seems most likely though that these forces would be used to extend Frontal Aviation's interdiction campaign deep into

become operational early in the decade, with some 100 being procured. This assumes that such an ambitious program does not run into serious technical problems. As the improved Soviet defenses will first be deployed around high priority targets and the B-1B is not entirely defenseless against new air defense systems, it is likely that the Stealth bomber will not completely supplant it in the penetration role until the mid-1990s.

enemy territory. Therefore the primary targets of Long Range Aviation's medium bomber force are likely to be airfields, ports, and other communications and supply centers. Secondary roles for LRA aircraft include intelligence-gathering and bolstering the Naval Air Force in supporting operations at sea.

Long Range Aviation's greatest contribution to the Soviet strategic armory at present is therefore as part of its theater nuclear forces, rather than as a supplement to the ICBM and SLBM forces targeted against the USA. Nevertheless a proportion of the bomber force could be employed on follow-up strikes over North America after an ICBM attack to pick out those targets missed by the missiles. It is significant that, despite the problems of operating a force largely composed of elderly bombers, the Soviet Union thinks it worthwhile to maintain LRA in its present form. The current missions of LRA could be as effectively performed by transferring its modern equipment to Frontal Aviation and the Naval Air Force. The Soviet Union has traditionally shown itself reluctant to retire any obsolete weaponry which has some remaining usefulness (this tendency can be seen throughout the Soviet armed forces). Yet even taking this into account the continued existence of LRA makes little sense unless the Soviet Union sees it as a cadre which can be expanded into an effective strategic bomber force once a modern intercontinental bomber becomes available.

Long Range Aviation is a subordinate command of the Soviet Air Force, equivalent in status to Frontal Aviation and Transport Aviation. It is commanded by a Colonel General of Aviation and enjoys a fair measure of operational independence. The basic operational unit of the force is the bomber regiment and these are assigned to three major commands. In the western Soviet Union, facing the NATO powers, are the Northwest and Southwest Bomber Corps, while the Far East Bomber Corps is deployed against China and Japan. The units of Long Range Aviation make use of air bases throughout the Soviet Union, including airfields in the Arctic region.

The current aircraft strength of LRA comprises some 800 aircraft, of which about 120 carry out the supporting roles of reconnaissance and inflight refueling. Inter-continental range bombing missions are assigned to some 100 Tu-20 Bear bombers and 45 Myasishchev M-4 Bisons, while the remainder of the force consists of medium range bombers. The most formidable of these is the Tu-26 Backfire, with more than 70 in service early in 1982 (plus as many with the Naval Air Force). This aircraft, although not a true inter-continental bomber, is able to attack targets in the USA on a one-way mission, landing at bases in Cuba. There are approximately 140 Tu-22 Blinders in service and the remainder of the medium bomber force consists of elderly Tu-16 Badgers.

The introduction of the Tu-20 Bear into Soviet service in 1955 followed by the M-4 Bison, resulted in a

dramatic expansion of the US air defenses. By the end of the 1950s USAF Air Defense Command controlled 40 regular fighter interceptor squadrons, in addition to ANG squadrons and an extensive radar network. In retrospect this hardly seems justified by a force of only some 200 Bisons and Bears, albeit nuclear armed.

Today seven USAF regular interceptor squadrons guard against a slightly smaller force of Soviet long range bombers. The Tu-20 is the largest Soviet bomber, spanning 167ft 8in and with a length of 155ft 10in. It is a swept-wing aircraft, powered by four massive Kuznetsov NK-12MV turboprops, each delivering 15,000shp (shaft horsepower). Top speed is 540mph and service ceiling 41,000ft. Its maximum range is 11,000 miles, which can be further extended by inflight refueling. Carrying a 26,500lb payload, the Bear's range is more than 7000 miles. In addition to free-fall bombs, the AS-3 Kangaroo stand-off weapon with a range of 400 miles, can be carried by some variants. A defensive armament of paired 23mm cannon is carried in dorsal and ventral remotely controlled barbettes and in a manned tail turret. Unlike the B-52, the Tu-20 does not appear to have been modified for the low level role and although a most impressive

Communications equipment aboard the E-4 national emergency airborne command post (above) enables it to communicate with strategic forces anywhere in the world. Apart from the bomber force, SAC also controls land-based ICBMs. Warheads from a Minuteman are seen reentering the earth's atmosphere (below).

aircraft in its day, the Tu-20 would present few problems to the present-day Western air defense system.

The Bear's stablemate is the swept-wing, turbojet-powered M-4 Bison, which serves both as a bomber and tanker aircraft with LRA. Powered by four 28,500lb thrust Soloviev D-15s, the M-4 attains a maximum speed of Mach 0.95 at 10,000ft and its service ceiling is 56,000ft. It can carry a 12,000lb payload to a range of 5000 miles and this can be extended by inflight refueling. Its warload consists of free-fall bombs only, as air-to-surface missiles (ASM) are not carried. The tanker version carries a drogue and hose unit on a reel fitted in the bomb bay. As with the Bear, the Bison's ability to penetrate modern air defenses is minimal and most of LRA's surviving M-4s may now be used as tankers. The Bison is fast approaching retirement and is likely to go out of service by the end of 1983.

The introduction of a modern Soviet intercontinental range bomber able to penetrate Western air defenses, both by flying at low level and by using ECM, is now long overdue. There have apparently been a number of attempts to fill this gap in the Soviet armory. In the late 1950s the Myasishchev bureau developed the M-50, a delta-wing, four engined bomber, with supersonic dash capability. However this aircraft never reached service, probably because it lacked the range for missions against the USA. More recently it was rumored that a bomber development of the Tupolev Tu-144 supersonic airliner was being tested. However this aircraft would be ill-suited to the high-speed, low level target penetration phase of the strategic bomber's mission and it is unlikely that the next generation Soviet strategic bomber will be this aircraft.

The most recent reports suggest that current Soviet strategic bomber development is following three lines. Firstly there are reports of a supersonic, variable-geometry bomber in the same class as the USAF's B-1. This would carry stand-off weapons and be equipped with a comprehensive range of ECM devices. The second reported development is of a subsonic, very long range multi-role aircraft, which could replace the Tu-20 Bear in its maritime roles, as well as providing a launch platform for stand-off weapons aimed at peripheral targets in North America. It is probably wrong to regard this aircraft as a cruise-missile carrier in the same class as the late-model B-52s. Although it is known to be developing such weapons, the Soviet Union is unlikely to produce an ALCM comparable with the American AGM-86B for some time. However, when a cruise missile capable of being launched outside the range of enemy air defenses does appear in the Soviet armory, it is likely that this new aircraft will be its carrier. Finally, the Soviet Union is believed to be working on an enhanced-range version of Backfire. If Backfire can be developed to attain an unrefueled range of 6000 miles, then it will be capable of true strategic missions against the United States.

The provision of a new tanker aircraft, while not presenting any of the formidable technical problems associated with the new strategic bomber, is also proceeding slowly. At present LRA relies on modified versions of the Bison and Badger for this necessary extension to the reach of its bomber force. Western sources believe that a tanker aircraft based on the Ilyushin Il-76 Candid transport is under development, but that its introduction has been delayed because of the pressing need for the military transport version. The wide-bodied Ilyushin Il-86 airliner has also been suggested as a possible basis for a tanker aircraft, on the lines of the McDonnell-Douglas KC-10 Extender version of the commercial DC-10. Whatever the future may hold for LRA, it is its medium range bomber force which at present represents the command's major contribution to Soviet military power. The spearpoint of this threat to Western Europe and the Far East is the Tu-26 Backfire bomber. This aircraft can operate as a theater strategic or tactical weapon, carrying nuclear or conventional weapons. There is some confusion over the Soviet designation of this aircraft, arising out of the SALT (Strategic Arms Limitation Talks) negotiations. The Soviet Union during these talks referred to Backfire as the Tu-22M (a designation that would logically be applied to a modified version of the Tu-22, which Backfire manifestly is not). I have therefore preferred the originally-identified designation Tu-26, which seems to fit the pattern of Soviet nomenclature more easily. Similarly, I have used the designation Tu-20 for Bear, rather than the currently-fashionable Tu-95 (most probably the design bureau's designation rather than the air force's) for the same reason.

The capabilities of Backfire have been somewhat distorted by the United States' understandable wish to have it classed as an inter-continental system during the SALT process. It is true that if the Soviet Union wished to employ Backfire on strategic missions against the United States this could be accomplished, although at high risk. With inflight refueling and making use of Arctic staging bases Backfire could attack many targets in the eastern and central USA and land afterward at airfields in Cuba.

The exact extent of its coverage of United States targets would depend on how much of the mission was flown at high altitude, where fuel consumption would be most economical but the risk of interception very high. It is also doubtful whether bases in Cuba would remain immune from attack by the United States in the event of war with the Soviet Union.

The threat of Backfire to the rear areas of NATO and to the vital Atlantic re-supply route is so great that Soviet commanders may well regard its use against US targets as wasteful. Backfire is the only modern Soviet warplane that can cover all of the United Kingdom, operating at low level throughout the mission (a lo-lo-lo-profile) from bases in East Germany. Flying from bases farther east, it could execute the same mission, using a hi-lo-hi profile, with the high level points of the flight over low-risk areas. It can also range over much of the North Atlantic from bases in the Murmansk region, and it could pose as great a threat to NATO shipping as the Soviet attack submarines. The initial deployment of Backfires supports this argument, in that deliveries have been divided equally between Long Range Aviation and the Naval Air Force. If Backfire was regarded as primarily a long-range strategic bomber, then deliveries to LRA would almost certainly have been given priority.

The production rate of Backfire was recently reported to have been accelerated from 2.5 to 3.5 per month, but if this is so it has yet to be reflected by the number of bombers reaching the aviation regiments. Although a highly capable aircraft, by Western standards Backfire is not very technologically advanced. Its variable-sweep wings, for example, are hinged at one-third of the span presumably to avoid the problems of trim change associated with the more efficient fully variable wings.

Span is 113ft with the wings unswept, reducing to 86ft in the fully-swept position and length is 132ft. Power is provided by two Kuznetsov NK-144 afterburning turbofans, giving 45,000lbs thrust each, and maximum takeoff weight is some 245,000lbs. Maximum speed varies from Mach 1.8 at 40,000ft to Mach 0.9 at sea level and range is some 5000 miles. Backfire is believed to carry a crew of four and has a defensive armament of two radar-directed 23mm cannon mounted in the tail. Its payload over a 5000 mile range is some 12,000lbs and weapons carried can include nuclear and conventional free-fall bombs, or stand-off missiles. Backfire has been photographed carrying one AS-4 Kitchen air-to-surface missile, but United States intelligence sources credit it with the ability to carry two of these weapons, or two of the more advanced AS-6 Kingfish ASMs. A 745 mile range cruise missile reportedly under development by the Soviet Union would presumably be carried by Backfire.

As originally conceived the Tupolev Tu-22 Blinder was to replace the Tu-16 with LRA, but only about 170 were produced for the command. The Tu-22 was a victim of changing operational conditions, rather than a failure in its intended role. It was designed to operate at high altitude and have a supersonic dash capability to defeat the NATO interceptors and SAMs of the late 1950s. However, by the time the Tu-22 reached the bomber regiments it was clear that speed and altitude were no protection against contemporary air defense systems. Consequently production was curtailed and the Tu-16 has soldiered on with LRA until well past its allotted lifespan.

The Tu-22 is primarily armed with the AS-4 Kitchen stand-off weapon, although it also carries free-fall nuclear and conventional bombs. Its operating radius in the strike role is about 1750 miles. Crew comprises the pilot and two other members seated in tandem behind him. Defensive armament consists of a

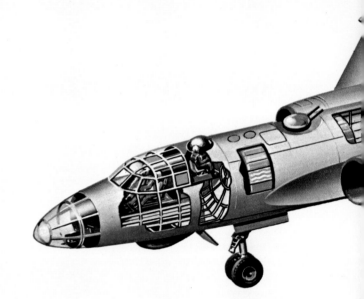

This cutaway view of a Tu-16 Badger, derived from Soviet sources, clearly shows the layout of crew positions defensive armament and weapons bay.

radar-directed 23mm cannon in the tail. The Blinder is a large aircraft for its class, its swept wing spanning 94ft 6in, overall length is 136ft 9in and maximum takeoff weight some 190,000lbs. Its maximum speed is Mach 1.5 at 36,000ft and service ceiling is 60,000ft. The engines, unusually mounted side-by-side high on the rear fuselage, are believed to be 31,000lbs thrust Kolesov VD-7 turbojets with afterburning. The ability of the Tu-22 to penetrate modern air defenses to any depth is extremely doubtful, but the bomber could be assigned to peripheral targets in Europe and it would fare much better over the People's Republic of China.

The mainstay of Long Range Aviation's theater nuclear capability (in terms of numbers of aircraft at least) remains the Tu-16 Badger. Although the type first flew 30 years ago, it still serves in its original medium-bomber role and also undertakes such tasks as inflight refueling, reconnaissance and electronic warfare. The Tu-16 can carry an 8000lb bomb load over a range of 3000 miles. The Badger-G variant carries two AS-5 Kelt ASMs over a 2000 mile range. Power is provided by two 19,200lb Mikulin AM-3M turbojets, giving the Tu-16 a maximum speed of 620mph, while service ceiling is 46,000ft. The Tu-16 carries a heavy defensive armament, comprising a fixed, forward-firing

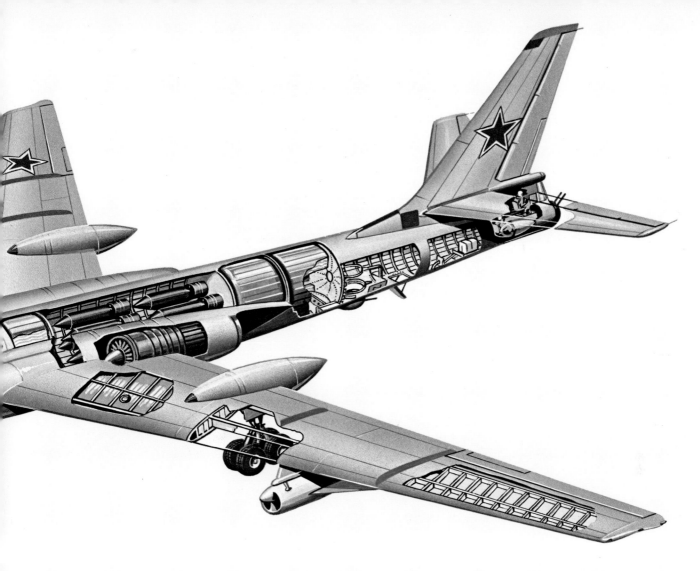

23mm cannon in the nose, with paired weapons of the same caliber in a manned tail-turret and remotely-controlled ventral and dorsal barbettes. Dimensions of the Tu-16 include a span of 113ft 3in and a length of 120ft and maximum takeoff weight is 158,000lbs. Although such an elderly bomber is obviously not capable of surviving for long in the high-threat environment created by modern air defense systems, it is likely to soldier on in less dangerous operational areas for some time to come. As the numbers of Backfire bombers build up to their anticipated total of 450 and the Su-24 Fencer takes over the interdiction/strike role in Frontal Aviation the Tu-16 will be phased out of service. The long life of the Tu-16 illustrates again the reluctance of the Soviet Union to discard any weapon so long as it has any useful life remaining. Perhaps it also reflects the effects of inter-service wrangling, with rockets competing with the manned bomber for defense funds which even in the Soviet Union are not limitless.

Although the two Superpowers hold a virtual monopoly of strategic nuclear forces, a number of second rank powers have a measure of nuclear capability. Britain has opted for a force of ballistic missile armed submarines at present armed with the American

Polaris missile fitted with British nuclear warheads. This is to be replaced by the American Trident missile. Until the early 1980s, the submarine force could be supported by six squadrons of Vulcan B Mk 2 bombers, which were part of Britain's strategic V-bomber force until superseded by Polaris in 1969. The Vulcans were thereafter retained in the strike/attack role, but by virtue of their long-range (some 4000 miles unrefueled at low level) and their ability to carry free-fall nuclear weapons, they are able to carry out theater strategic missions. With the introduction of the Panavia Tornado into front-line service with the RAF in 1982, the Vulcan squadrons were rapidly run down. The Tornado GR Mk 1 is primarily an interdictor/strike aircraft, but like its predecessor it can carry free-fall nuclear weapons and has the range to cover targets in the western part of the Soviet Union.

France's nuclear forces are divided between land and submarine based missiles and manned bombers. The bomber element of the Forces Aériennes Stratégiques comprises nearly 50 Mirages IVs, supported by eleven C-135F tanker aircraft. The bombers currently serve with the 91e Escadre de Bombardement with headquarters at Mont-de-Marsan and the 94e Escadre de Bombardement at Avord. Other bases are used by the component *escadrons* (of four aircraft each) to provide a degree of dispersal for the bombers. The C-135F tankers belong to the 93e Escadre de Ravitaillement en Vol with headquarters at Istres. Training is carried out by the Centre d'Instruction des Forces Aériennes Stratégiques No 328 at Merignac, which also carries out strategic reconnaissance with four Mirage IVAs fitted with CT-52 reconnaissance pods. The first production Mirage IVA first flew in December 1963 and delivery of 62 aircraft was completed in March 1968. A tail-less delta, like the smaller Mirage III fighter, the Mirage IVA spans 38ft 10in and length is 77ft 1in. Power is provided by a pair of SNECMA Atar 09K-50 turbojets, each delivering 15,400lbs thrust with afterburning. Maximum speed is Mach 2.2 at 60,000ft and service ceiling 65,000ft. Unrefueled range is 2500 nautical miles. The crew of two, pilot and navigator, are seated in tandem. In the strategic role a 60 kiloton free-fall nuclear weapon is carried semi-recessed in the underside of the fuselage. A conventional bomb load of 16,000lbs can be carried, or four AS37 Martel anti-radiation missiles, which operate by homing on enemy radar emissions.

Originally conceived as a high-level penetration bomber, the Mirage IVA has been modified to undertake low-level sorties. The importance of the manned bomber in the deterrent role has diminished with the build-up of the Navy's missile-armed submarine fleet. However the ASMP (air-sol moyenne portée) stand-off nuclear missile to be fitted to the aircraft from 1986 will extend its useful life into the 1990s. The ASMP's 60-mile range eases the task of target penetration and provides an interim theater nuclear system until the introduction of the Mirage IVA's successor. This is apparently not to be another manned bomber, despite the suitability of the privately-developed Mirage 4000 for this role. The favored solution is a French-developed long-range cruise missile, or alternatively a mobile IRBM (intermediate range ballistic missile) with multiple warheads similar to the Soviet SS-20.

The People's Republic of China exploded its first nuclear weapon in 1964 and various strategic delivery systems have been developed. These include land-based ICBMs, IRBMs and MRBMs (medium range) and a submarine-launched system also. The primary manned bomber system is the B-6, a Chinese version of the Tu-16. Some 60-80 of these are believed to be in service and the type may still be in production. A backup nuclear bomber may be the Tupolev Tu-4, a Soviet copy of the World War II-vintage Boeing B-29 Superfortress, which surprisingly still serves in small numbers. These aircraft are of course targeted against the Soviet Union, but their ability to penetrate Soviet air defenses must be extremely limited. They are most probably regarded as interim systems, pending the development of fully-reliable Chinese ICBMs and SLBMs. However, it should be borne in mind that even a highly-efficient air defense system will be fortunate to

Minuteman III's three warheads (above) yield 165 kt, while Titan II's single warhead (right) yields 10mt.

achieve a 100 percent interception rate and that any nuclear armed bomber reaching its target will inflict a very high level of damage. It is therefore unwise totally to discount the potential dangers posed by the Chinese strategic forces.

Discussion of strategic bomber forces not unreasonably centers on their nuclear capabilities. Yet it is worth remembering that they are able to carry out conventional strategic operations with high explosive bombs. This conventional capability also makes the strategic bomber a useful tactical weapon, as was illustrated by the long series of Arc Light sorties by B-52s over South Vietnam during the Southeast Asia conflict.

The B-52 can also play a useful role in a limited naval war, either with bombs and stand-off weapons, or as an aerial minelayer. The Soviet Union similarly sees an important maritime role for the bombers of LRA. The great problem with employing strategic forces for tactical purposes is the risk of losing strategic bombers in secondary operations. Therefore such tactical employment tends to be assigned to the obsolescent bombers in the force. Conventional strategic operations are rather different, in that the objective justifies a greater degree of risk. The Linebacker II campaign against North Vietnam in 1972, which resulted in the resumption of the Paris Peace Talks, falls into this category. It is possible that similar actions could be mounted early in a major conflict.

A major factor affecting the design of strategic bombers and the development of their tactics is the need to evade hostile air defenses to reach the target. This requirement led to the introduction of low-level penetration tactics when medium and high altitude SAMs and interceptors made any other flight level unsafe for the bomber. The radars and communications which direct enemy interceptors and missiles can be jammed or deceived by electronic countermeasures. Flares can be released to decoy heat-seeking infra-red missiles away from the bomber's jetpipes and radar-guided missiles can be similarly foiled by clouds of chaff. At one time SAC's B-52s carried the ADM-20 Quail, a miniature decoy aircraft that could be released from the bomber and present enemy radars with the same return signals as its parent. Ground based defense systems can be directly attacked by SRAM missiles and interceptors can be engaged by the bomber's defensive gun armament.

The stand-off weapon, or ASM, is one of the most effective means of attacking a heavily defended target area, as the attacking aircraft does not have to overfly its objective. Among the Soviet ASMs in service with LRA is the elderly AS-3 Kangaroo, which is carried by the Tu-20 Bear. This large missile, with a span of about 29ft and a length of 49ft, is in effect an unmanned, swept-wing fighter aircraft powered by a turbojet and guided by autopilot. It carries a nuclear warhead and has a range of some 400 miles. Accuracy is doubtful against all but the largest targets and the AS-3's cruising speed of Mach 2 would not provide immunity from interception.

The AS-4 Kitchen is a more advanced ASM, which arms the Tu-22 Blinder and Tu-26 Backfire. It is believed to be powered by a liquid rocket motor and cruises at Mach 2 over a range of about 200 miles. Guidance is believed to be inertial and a nuclear warhead is probably fitted. Information on the later AS-6 Kingfish is even more sparse, but it is thought to have a considerably improved performance and accuracy over the AS-4. It is inertially guided to its target and may have a terminal homing system. Cruising speed is Mach 3 over a range of up to 400 miles. The warhead yield is thought to be 200 kilotons.

The Boeing AGM-69A SRAM was the only stand-off weapon in service with SAC until late 1982. It is primarily intended as a defense suppression missile, punching a hole in the enemy air defenses which can then be penetrated by the bomber carrying its free-fall weapons. It carries a 200kt thermonuclear warhead, is powered by a solid-propellant rocket motor and has inertial guidance. SRAM is an agile and versatile missile – it can be programmed to follow a semi-ballistic flight path to its target, it can follow the terrain at

The USAF's new MX missile carries ten independently-targetable warheads and is due to enter service with SAC's strategic missile wings in the late 1980s.

low-level, or pull up from behind shielding terrain to dive onto its target. Its range depends on the way it is employed, but the maximum is a little over 100 miles. Speed is over Mach 3. Typical targets for SRAM would be Soviet air defense airfields, early-warning radar sites, control centers and SAM batteries. It could also be employed against primary strategic targets.

A longer-ranging stand-off weapon than SRAM is now entering the strategic lists, after the Boeing AGM-86B ALCM entered service with the USAF's 416th BW at Griffiss AFB, NY during 1982. The cruise missile is by no means a new concept, as it can trace its ancestry back to the VI (Fieseler Fi 103) of World War II. During the 1950s the same idea produced such weapons as the US Navy's Regulus and the USAF's Matador, Mace and Snark, none of which was very successful.

There are three main reasons for the reappearance of the cruise missile, not only in its air-launched version, but also as a ground launched weapon and submarine and surface ship armament. The first is the development of TERCOM (*terrain contour matching*) an advanced navigational system, accurate over long distances. Second is the advance in powerplant technology, which makes a small light turbofan with economical fuel consumption a practical proposition. Finally there is the work of the nuclear physicists, who have produced a miniature nuclear warhead of high yield (200 kilotons). Soviet progress in these fields lags behind that of the United States and a Soviet equivalent to ALCM is some way off.

The Boeing AGM-86B ALCM is shaped like a miniature swept-wing aeroplane, spanning 12ft and with a length of 20ft 9in. It is powered by a 600lbs thrust William Research Corp F-107 turbofan, which gives a cruising speed of more than 500mph. The triangular-section fuselage houses the guidance system and warhead in the forward section, with the powerplant to the rear and the fuel tank occupying much of the space between. The low-mounted 25 degree swept wing folds under the fuselage when the ALCM is mounted on its parent aircraft.

During its initial operational deployment the ALCM will be carried on the B-52G's wing pylons only. This gives a maximum load of 12, which will be increased by eight missiles when ALCM is carried by the internal rotary launcher. The B-1B's load of ALCMs is 14 carried externally and eight mounted in the weapons bay rotary launcher. A constraining factor in the original ALCM program was the need to accommodate the missile within the length of the B-1A's internal weapons bay. This restricted the size of the missile and consequently its range. However, with the cancellation of the B-1A this limitation was removed and the AGM-86A was superseded by the longer-ranging AGM-86B. With a range of some 1500 miles, this ALCM can attack some 85 percent of strategic targets in the Soviet Union when launched from a point beyond the effective coverage of the Soviet air defense

The cruise missile depends on a combination of its small size (and therefore inconspicuous radar signature, which can be yet further reduced by Stealth techniques) and low operating altitude to evade enemy defenses. Unlike the manned bomber, it does not make use of ECM to degrade enemy air defense systems, although developments in automation and miniaturization may make this possible in the future. However, even if the Soviet Union develops radar and missiles with sufficiently accurate lookdown/shootdown characteristic to cope with these small targets, the defenses will be in danger of becoming swamped by sheer numbers. The cruise missile requirement for the B-52 force alone is estimated at 3400 rounds.

At the time of the B-1A's cancellation, the cruise missile was presented as a wonder weapon in comparison with the manned bomber which it would replace.

A Minuteman ICBM (above left) is loaded into its launch silo by crane. Operational missiles are carried to their silos in special transporters.

umbrella. The need to cover the remaining 15 percent of strategic targets, and therefore for the launch aircraft to be capable of penetrating Soviet air defenses, has resulted in a cruise missile armament for the revived B-1B. The problem of accommodating the larger ALCMs in the bomber's weapons bays has been met by inserting a removeable bulkhead between the two forward bays, allowing them to be converted into a single bay long enough to house the AGM-86B.

Up until 1980 the Boeing AGM-86B was in competition with the General Dynamics AGM-109 Tomahawk to meet the USAF's ALCM requirement. If the Tomahawk had been selected, the US services would have standardized on this weapon, as it will become the US Navy's cruise missile and the USAF's GLCM (ground-launched cruise missile) as part of the theater nuclear force modernization program. However, the Boeing ALCM was preferred for the SAC mission on the grounds of its superior guidance system and flight characteristics, although in terms of overall performance the two systems are largely comparable.

The penetration capability of the latter was considered dubious whereas the cruise missile's ability to avoid interception was unquestioned. Wiser counsels have since prevailed and, with the rediscovery of the bomber's virtues, has come a recognition of the ALCM's limitations. As with ballistic missiles, they lack flexibility and cannot be retargeted after launch. Furthermore they will be unable to recognize and respond to threats from air defense systems. They do not have the speed of the ICBM (a flight time of under 30 minutes, compared with eight hours or more for ALCM), nor can they carry the large weapons load of a bomber. This is not to suggest that the advantages of the ALCM are illusory, but rather that it is a complement to existing systems; certainly not a weapon against which there is no defense.

After its launch at a distance of some 200 miles from

the outer edge of the enemy air defense zone, the ALCM will cruise at an altitude of 5-10,000ft before descending to a terrain-following height of around 150ft. Guidance is by means of an inertial system, which is accurate over a short flight but will develop unacceptable errors during a typical ALCM mission. It is therefore periodically corrected by means of the TERCOM technique. This compares the terrain over which the missile flies with stored data, enabling course corrections to be made to bring the ALCM back to its precomputed flight path. Thus the ALCM can be guided to its target with great precision, as the degree of accuracy can be increased at every update of the navigational information. The final phase of the ALCM's flight is assisted by a scene-matching area correlator (SMAC). This carries information on the geography of the immediate target area, which is compared with that

A liquid-fueled Titan II blasts from its launch silo at Vandenberg AFB, California, where all USAF missile test firings take place.

picked up by the ALCM's sensor and necessary last-minute corrections are made to the missile's flight path. Accuracy of the ALCM is such that CEP (circular error probable) is measured in tens of feet.

At one time suitably-modified wide-bodied civil airliners were under consideration as cruise missile carriers. However it was finally concluded that the cost of modification would outweigh the advantages of using an existing design. Nor would such an aircraft have any ability to penetrate enemy air defenses. A similar, technically feasible, proposal to launch the MX missile from a modified transport aircraft is likely to be rejected in favor of one of the land-based modes for this controversial ICBM.

Before considering the land-based ballistic missiles which form such an important part of the strategic armory, airborne command, control and communications (C^3) deserve some mention. Essentially airborne command posts are back-up systems for SAC's land-based installations, which may be knocked out by missile attack. The aircraft used for this duty are EC-135 variants of the KC-135 tanker. Since 1961 a relay of EC-135 airborne command post aircraft has been continuously in the air carrying a battle staff under the command of a general officer. Named Project Looking Glass, this airborne command center is in communication with all the important national headquarters and with SAC bomber and missile forces. It can take over the control of strategic forces and order them into action, indeed Minuteman missiles can even be fired by controllers aboard the aircraft. EC-135s can also pro-

vide the basis for a Post Attack Command and Control System for forces surviving a nuclear attack. Similar aircraft also provide command posts for the Joint Chiefs of Staff, C-in-C Europe, C-in-C Atlantic and C-in-C Pacific.

A system of greater capacity is the E-4B national emergency airborne command post, which may carry the President of the United States. The E-4 airframe is essentially that of the Boeing 747 airliner, but it is fitted with an extensive array of equipment to suit it for the demanding role. This ranges from super-high frequency communications with satellites to very low frequency (VLF) radio links to submerged submarines. The E-4B can remain in the air for up to 72 hours, assuming that inflight refueling is available. Two complete flight crews will be carried and the command staff can number as many as 50 people. Four of these aircraft are to be based at Andrews Air Force Base, Maryland, close to Washington DC.

The US Navy makes use of specially modified Lockheed Hercules (EC-130G and EC-130Q variants) as communications relays to its SLBM-armed submarine fleets. These aircraft carry the Collins TACAMO (take command and move out) VLF radio systems, which can relay orders to submerged submarines. This is essentially a back-up system for shore installations, but it is considered sufficiently important to justify having at least one aircraft continually available. The

EC-130Q is operated by Navy squadron VQ-3 based at Agana, Guam, in the Pacific. The Atlantic is covered by VQ-4, which flies EC-130Gs and EC-130Qs from Patuxent River, Maryland.

The firmest leg of the US strategic triad has long been considered to be the land-based ICBM. This weapon has enjoyed several important advantages over the SLBM and the bomber/cruise missile. It is more accurate than present SLBMs and does not have the same problems of C^3. Its flight time is considerably less than that of the bomber and its effectiveness should not be dependent on advanced warning of enemy attack. Furthermore present missile systems are very reliable, with few vehicles unavailable because of unscheduled maintenance requirements. However, with the long-delayed deployment of MX (scheduled to be fully deployed by 1989) a 'window of vulnerability' has been judged to have opened over the US ICBM force. This is because the Soviet Strategic Rocket Forces are claimed to be fast developing the capability of knocking-out the US missile force in a pre-emptive first strike.

SAC's current ICBM force comprises 1000 Boeing LGM-30 Minuteman missiles and 53 elderly Martin LGM-25C Titan II missiles. They are assigned to nine strategic missile wings (SMW) which are deployed as follows:

Unit	Base	Composition
44th SMW	Ellsworth AFB, SD	3 squadrons of LGM-30F Minuteman II
90th SMW	FE Warren AFB, Wy	4 squadrons of LGM-30G Minuteman III
91st SMW	Minot AFB, ND	3 squadrons of LGM-30G Minuteman III
308th SMW	Little Rock AFB, Ar	2 squadrons of LGM-25C Titan II
321st SMW	Grand Forks AFB, ND	3 squadrons of LGM-30G Minuteman III
341st SMW	Malmstrom AFB, Mt	3 squadrons of LGM-30F Minuteman II 1 squadron of LGM-30G Minuteman III
351st SMW	Whiteman AFB, Mo	3 squadrons of LGM-30F Minuteman II
381st SMW	McConnell AFB, Ks	2 squadrons of LGM-25C Titan II
390th SMW	Davis-Monthan AFB, Az	2 squadrons of LGM-25C Titan II

The Titan II is the only survivor from the early generation of liquid-fueled American ICBMs, which included the Titan I and Atlas D, E and F. Unlike the first ICBMs, which had to be fueled immediately before launch, the Titan II propellants can be stored within the missile and only deteriorate slowly. It is SAC's largest missile, standing 130ft high, with a launch weight of 330,000lbs. It also carries the largest warhead in the US armory with a ten megaton yield. Its first-stage rocket motor produces 430,000lbs thrust and the second stage, which ignites at an altitude of 250,000ft delivers 100,000lbs of thrust. The guidance system is inertial and the missile is housed in a 150ft deep silo.

For many years the Titan II force remained constant at 54 missiles, but in September 1980 this was reduced by one in a much publicized accident at Little Rock AFB, Arkansas. Each Titan II squadron comprises nine missiles, with each silo having its own control center manned by four crewmembers. This missile is to be phased out of service as part of the modernization program announced in October 1981.

The present Minuteman force is made up of two models. The LGM-30F Minuteman II carries a single thermo-nuclear warhead, whereas the LGM-30G Minuteman III has three multiple, independently-targeted re-entry vehicles (MIRVs). There are 450 Minuteman IIs in service and 550 Minuteman IIIs, but the latter force may be reduced as MX comes into service to keep within the warhead limits imposed by SALT.

Both models of Minuteman stand 59ft 10in high and have a maximum diameter of 6ft. Minuteman II weighs 70,000lbs at launch, and Minuteman III is 6000lbs heavier. Both are three-stage missiles, with solid propellant rocket fuel. However Minuteman III has a larger third-stage and its post-boost guidance system associated with the MIRV is powered by a liquid-fueled rocket. Range of Minuteman II is 7000 miles and Minuteman III's range is 8000 miles. Minuteman II's warhead has a yield of between one and two megatons and Minuteman III's warheads yield 165 kilotons each. Minuteman II became operational in 1965 and was joined by Minuteman III five years later.

Minuteman silos, 80ft deep, are clustered in groups of ten, which are controlled from a single center crewed by two officers. A separation of at least three miles is needed between each silo and also the launch control center. A number of improvements to the missile system are designed to enhance survivability and improve accuracy. They include further hardening the silos by reinforcing their covers with boron-impregnated concrete, improving the suspension system to reduce damage from shock waves and fitting bins to silo doors to catch debris from nearby explosions which might otherwise jam the doors shut. Command and control improvements allow remote retargeting of the Minuteman within 30 minutes and some missiles can be retargeted from airborne command posts. The missile's electronics are being further hardened against damage from EMP. Finally a number of Minuteman IIIs are to be refitted with Mk 12A warheads of 330 kiloton yield.

It is argued that the impending threat to the Minuteman from Soviet ICBMs, makes the speedy deployment of the new MX missile (and a decision on its basing mode) a matter of the utmost urgency and importance. Yet the attitude of President Carter's administration and that of his successor to a lesser degree, does little to reflect the perception of a serious threat to a major element of the United States' strategic forces.

The basing of the MX missile has generated more controversy than the technical characteristics of the weapon itself. Obviously the major requirement for a new basing method is that it does not present the tempting target that the silo-based Minuteman force will do in the near future. Various methods of achieving this have been discussed, but one complicating factor has been the requirement that the strength of the MX force can be verified by Soviet satellite reconnaissance to further the SALT negotiations. To a large degree the requirements of survivability and verification are incompatible, because the best means of ensuring that MX can survive a pre-emptive strike is by concealing and dispersing the missiles.

One of the possible basing modes favored by strategic planners, the multiple protective structures (MPS) mode, involves the construction of a cluster of protected, horizontal shelters between which the missiles would be shuttled at random. Under present plans, for every shelter that contained an MX missile, there would be 22 empty shelters which would have to be targeted by Soviet missiles to ensure an effective first strike. This clearly presents the Soviet forces with an impossible objective. The MPS plan is opposed by environmentalists, who fear its effects on local communities in the Nevada/Utah area where the shelters would be built.

A less objectionable plan, but one which would take longer to implement, involves building super-hardened silos in closely grouped clusters. Missiles could be moved from silo to silo giving a degree of dispersal and the close grouping of silos would paradoxically also give protection. This is because a warhead exploding on one silo would tend to knock warheads targeted against adjoining silos off course, a phenomenon known as 'fratricide'.

Alternative basing modes for MX have serious technical objections or operational shortcomings. For example it has been suggested that MX be based on small diesel-electric submarines operating in the shallow waters off the US east coast. However such a force would be very vulnerable to the shock waves from a nuclear explosion (which need not be precisely targeted on the MX-carrying submarine). Air mobile MX missiles were also considered and in 1974 a Minuteman missile was successfully launched from a Lockheed C-5A transport aircraft. However, this mode relies on timely warning of attack to be fully effective and would be costly to develop and maintain.

The active defense of MX sites by ballistic missile defense (BMD) systems is not feasible in the short term. However, progress in this field may enable a limited system to be deployed by 1990. If effective BMD could be deployed sooner, it would remove the Soviet threat to the Minuteman silos. As it is, the remaining alternatives, assuming that the approaching Soviet capability to take out the Minuteman force has been correctly assessed, are to develop an MPS system for this missile, or to launch the force on warning of attack. Cost and timescale of the first alternative effectively rule it out as a practical policy. Launch on warning of attack is unsatisfactory because at best the warning time will be only some 20 minutes and the indications of attack may be ambiguous. However until the deployment of MX, the US President may well have to make this awesome decision under the crude but compelling logic that if you are going to lose it, then you should use it first.

The discussion of the vexed question of MX basing tends to obscure the technical merits of the missile itself, a major advance over the 1960s technology of Minuteman III. It is a three-stage missile, with a fourth post-boost stage to achieve final positioning of the re-entry system. MX carries no fewer than ten Mark 12A MIRVs, each with a 330 kiloton yield warhead. A mobile MX system would include a launcher to push the missile from its shelter and erect the missile canister for launch. A wheeled transporter would move the missile by road from the assembly area to the operational site and between shelters. At present 100 MX missiles are on order for delivery by 1989. The first of these could be placed in existing Minuteman silos, pending a final decision on basing.

Illustrations of Soviet strategic missiles are rare. An ICBM is seen in its launch silo (below), while an SS-4 medium-range missile lifts off (right).

The USSR, like the USA, regards its ICBM force as the most important weapon in its strategic armory. Much effort has been expended on building up the ICBM force, both qualitatively and in numbers and by the early 1980s it was generally agreed that the Soviet Union had achieved strategic parity with the United States.

By the beginning of 1982 the Soviet Union had deployed 1398 ICBMs, of which over half were modern, fourth-generation missiles (SS-17s, SS-18s and SS-19s), the remainder being the older, third-generation SS-11s and SS-13s. The numbers in service are as follows:

Missile	Nos Deployed
SS-11	580
SS-13	60
SS-17	150
SS-18	308
SS-19	300

The SS-11 Sego (these identifications are NATO reporting names not Soviet designations) has been in service since 1966 and was deployed in large numbers during the 1970s. A program is underway to convert SS-11 silos to accommodate the SS-19 Mod 3, but it is likely to take several years to complete. SS-11 is a two-stage missile fueled with storable liquid propellant. It has a launch weight of some 106,000lbs and its range is an estimated 6500 miles. The original version could carry a single warhead, either of 500 kiloton yield or alternatively in the 20-25 megaton range. The later SS-11 Mod 3 carried a multiple re-entry vehicle (MRV) with three 300 kiloton warheads, although these were not independently targetable but fell in a pre-determined pattern.

The SS-13 Savage is believed to be the first Soviet ICBM to have solid-fuel rocket motors. It is a three-stage missile with a launch weight of some 77,000lbs. Range is around 5000 miles and a single warhead of one megaton yield is carried. SS-13 entered service in 1968 and a successor, the SS-16 was developed but never deployed.

Together with the SS-19, the SS-17 ICBM is a successor to the SS-11 and has been mounted in former SS-11 silos. It has a launch weight of some 143,000lbs and a range of more than 6000 miles. It is a two-stage liquid fueled rocket which is 'cold launched', which means that the missile is ejected from its silo by compressed gas and the rocket motor ignited above ground. This technique minimizes damage to the silo during launch and enables it to be rapidly reloaded with a second missile. The SS-17 Mod 1 carries four MIRV warheads, but most deployed missiles are SS-17 Mod 2s with a single high yield warhead. This suggests that SS-17 has sufficient accuracy to be used in a counter-force role against US ICBM silos.

The largest missile in the world, SS-18 combines a high payload with a greater accuracy than earlier Soviet missiles. It is a two-stage, cold launched missile, with a launch weight of around 485,000lbs and a range of 7500 miles. Four variants have been identified, Mod 1 having a single warhead of 24 megaton yield and a CEP of 1300ft. Mod 2 has eight to ten MIRVs and CEP is the

The US Army's Pershing I battlefield nuclear delivery system is now due for replacement by longer-range Pershing IIs and Ground Launched Cruise Missiles.

same as Mod 1. Mod 3 has a greater range than earlier SS-18s and improved accuracy (CEP is 1250ft). Mod 4 shows a further improvement in range, carries ten 500 kiloton MIRVs and has a CEP of 850ft.

SS-19 is reckoned to be the most effective of the fourth-generation Soviet ICBMs. It is a two-stage liquid-fueled rocket with a launch weight of 172,000lbs and a range of more than 6000 miles. Three versions have appeared in service, but the Mod 3 appears to be definitive and earlier missiles are being modified to this configuration. Little is known of the Mod 3's warhead, except that it is made up of MIRVs. The earlier Mod 1 carries six 550 kiloton MIRVs and the Mod 2 a single warhead. Two new Soviet ICBMs are reported to be approaching the flight test stage and could be deployed by the mid-1980s. One is an SS-17-class missile to be deployed in super-hardened silos. The other is mobile and of similar size to MX. Both use solid fuel. A third development is believed to be a large liquid-fueled ICBM in the same class as the SS-18.

One of the lesser nuclear powers, China, has developed ICBMs. However, their operational status is doubtful, but the limited range (3500 nautical miles) CSS-3 is believed to be in an advanced state of development and some sources claim a few are now in service. The later CSS-4 is a true ICBM with 6000 mile range and a five megaton yield warhead. The CSS-4 can reach targets throughout the USA and some may be operational. Yet the Soviet Union is China's chief potential antagonist and the CSS-3 has sufficient range to cover any Soviet target.

Another element in the strategic equation which is sometimes overlooked is those parts of the theater nuclear forces which are theater strategic rather than tactical nuclear systems. However the distinction is not a clearcut one and many weapons can operate in both roles. For example the Soviet SS-4, SS-5 and SS-20 IRBMs and Tu-16, Tu-22 and Tu-26 bombers fall into this category. Similar NATO weapons include Ground Launched Cruise Missile (GLCM) and Pershing II, neither of which is yet deployed, and F-111, Tornado and Jaguar strike aircraft.

SS-20 is a formidable mobile missile, which carries three MIRVs and can cover any NATO target in Western Europe. More than 260 had been deployed by the beginning of 1982 according to US official sources. NATO wishes to counter this weapon with the Pershing II missile and GLCM. It is feared that the Soviet Union may be gaining a local superiority in theater nuclear forces which would deter NATO commanders from using their tactical nuclear weapons to blunt a Soviet attack for fear of Soviet retaliation in kind against strategic targets in Europe. The modernization of NATO's theater nuclear forces would restore the nuclear balance within the European theater and remove the unacceptable constraints on NATO's defensive plans which a Soviet superiority in 'Euro-strategic' weapons could create.

The US Space Shuttle pictured during an early test flight atop a modified Boeing 747. Because the Shuttle can be reused many times, it represents a considerable advance in space operations.

2. WAR IN SPACE

War in Space

On 1 September 1982 the USAF formed Space Command, with headquarters at Colorado Springs, to co-ordinate all USAF space missions. This reflects the ever increasing use of space for military purposes. In the 25 years since the first successful space launch – the Soviet Union's Sputnik 1 satellite in October 1957 – the military capabilities of space craft have steadily expanded. Satellites are now used for reconnaissance, surveillance, ballistic missile early warning, electronic intelligence gathering, communications, weather reporting and navigation. Manned space missions too have their military applications. Two of the five Soviet Salyut space stations launched into earth orbit were used exclusively for military purposes and 118 of the 312 Space Shuttle missions expected up until 1994 will be for the Department of Defense. Nor are the military uses of space entirely passive, as both the United States and the Soviet Union are actively pursuing the development and deployment of anti-satellite weapons and space-based ABM systems are also in prospect.

Reconnaissance satellites are by far the most numerous and highly developed military space systems at present. Since 1960, when a Soviet SAM brought down the Lockheed U-2 reconnaissance aircraft flown by Francis 'Gary' Powers, manned overflights of the Soviet Union have involved unacceptable risks. However, satellite reconnaissance was swiftly to take over the vital task of strategic reconnaissance over Soviet territory. The USA launched six reconsats in 1960 and this figure had increased fivefold by 1962. By the early 1970s the annual launches had dropped to under ten, reflecting the greater versatility of the later American reconsats.

In the early years of satellite reconnaissance the United States had used two basic types: one for area reconnaissance and the other for 'close look' missions. The latter type would cover such locations as the Tyuratam complex in Soviet Central Asia where new ICBMs are tested and the air force's experimental test center at Ramenskoye. The features of these two satellites were combined into a single system, with the advent of Project 467 or 'Big Bird' in 1971. This massive vehicle, which is nearly 50ft in length, carries two cameras: an Eastman Kodak for area survey and a Perkin-Elmer high resolution camera for detailed coverage. It can also carry other reconnaissance sensors, such as side-looking radars, and is able to undertake infra-red photography.

Big Bird follows a slightly higher orbit than the close-look satellites and has a typical operating life of five months. The satellite carries six recoverable capsules, which can be ejected and returned to earth with exposed film. These are recovered over the Pacific by modified NC-130B and NC-130H Hercules aircraft, fitted with a special tracking radar and mid-air snatch aparatus, which operate from Hickam AFB, Hawaii. Ground stations can transmit commands to the satellite

The US maintains a continuous watch on space. This is the ballistic missile early warning station at Thule, Greenland.

for alterations in orbit, which are achieved by firing a boost motor in short bursts, and Big Bird can relay telemetry back to earth.

The latest American reconsat, KH-11, has taken this process a stage further and is able to transmit its pictures back to earth, rather than return them in a capsule. One outcome of this development is a greatly extended life for the satellite and KH-11 is able to operate in space for about one year. Another advantage is that, because the photographs are interpreted soon after they are taken, they become a source of short-term intelligence whose value may only last for a week or so. However, the advantage of timely information is balanced by a loss in picture quality and so KH-11 has supplemented rather than replaced Big Bird. A number

of 'close look' satellites also remain operational and these are launched to observe events of special interest. An example of this was the observation of the Soviet armed forces' reaction to the rioting in Poland in 1982.

The Soviet Union has launched many more reconsats than the USA, the yearly average being around 35. They fall into two categories of mission. One lasts for about a fortnight and returns film at the end of the mission only. This type is used for area coverage and indeed very similar missions are flown for earth resources surveys of the Soviet homeland. The second category produces high resolution, close-look coverage of important military installations. These missions can last up to 44 days, with film being returned twice during the mission and at its end. As many as four reconsats may be operational at one time and their coverage can be supplemented by photography and direct observation from manned spacecraft.

The results of military space reconnaissance missions are highly classified and no pictures taken by reconsats have ever been released for publication. However, it is known that the resolution of such pictures is high and images can be enhanced by computer-assisted techniques during interpretation. It has been claimed that it is possible to identify a Soviet officer's rank insignia on his shoulder boards from a reconsat photograph. Whatever the truth of this, it is certain that a great deal of detailed information can be gleaned from satellite photographs. The West owes its knowledge about the new generation of Soviet warplanes to this source and indeed the 'Ram' designation which they are given derives from the airfield (Ramenskoye) where they are first photographed. Intelligence about new developments and deployment of strategic missiles comes from satellite reconnaissance and both Superpowers are so confident of this source that they

Space satellites perform numerous important military roles ranging from strategic early warning to reconnaissance. The satellite pictured (above left) is Skynet, a British military communications system purchased from the US. The control center of US meteorological satellite programs (above) is at Fairchild AFB, Wa. The placing of satellites in orbit will be greatly facilitated by use of the Shuttle (left).

require no further verification that SALT ceilings are being observed.

Satellites are not limited to visual reconnaissance and much information which is otherwise unobtainable can be gained by electronic intelligence (ELINT) or 'Ferret' satellites. These are equipped to monitor and record radio and radar transmissions and, unlike ELINT aircraft operating outside the Soviet frontiers and the NATO ground stations in Turkey and those in the People's Republic of China, they can penetrate to the very heartland of the Soviet Union. ELINT satellites can 'capture' the telemetry transmitted from Soviet missiles under test, so that after this is decoded the United States has the same performance data as the Soviet Rocket Forces. Ferrets can also detect and classify radar transmissions, so that a picture of the Soviet air defense ground environment can be built up. Electronic eavesdropping is also useful in pinpointing military headquarters. The American ferret satellites are launched riding 'piggy back' on Big Bird, before separating to follow a high circular orbit. The Soviet Union also uses satellites for ELINT, with a network of six in circular earth orbit. These have a lifetime of about 18 months and are invariably replaced before failures occur.

The advent of the cruise missile has led to a new specialized reconnaissance role for satellites. The cruise missile's TERCOM system requires very accurate information on the physical geography of Soviet territory which can only be obtained from satellite ground mapping. In the course of gathering this data, it was discovered that the positions of Soviet cities as marked on published maps were inaccurate by several miles.

Weather reconnaissance is one of the most familiar applications of non-military satellites, but it is also of great value to the armed forces. For this reason the USAF has its own weather reconnaissance satellite system, the information from which is collated by the Global Weather Central, located with SAC headquarters at Offutt AFB, Nebraska. SAC is of course an important recipient of meteorological reports, which can cover any part of the earth. However, it is far from being the only user and satellite weather pictures can be transmitted to army units in the field and aircraft carriers at sea. If tactical nuclear weapons are used on the battlefield, metsats will plot the movement of plumes of radioactive fallout, so that friendly troops knew which areas are free from contamination. Similar information can be provided if chemical weapons are employed. Conversely, meteorological data is also used to plan the use of nuclear, bacteriological or chemical (NBC) weapons by one's own forces to the best effect. Lastly metsats are useful in the control of reconsats, by giving forewarning of cloud cover over an area to be photographed.

Ocean surveillance is one role in which satellites can operate most effectively, covering areas which would require very many aircraft to survey. The US Navy's White Cloud program aims to provide the position of every Soviet surface ship. This information has become of particular importance with the expansion of the Soviet Navy in the 1960s from little more than a coastal defense force into a true ocean-going navy with base facilities in many parts of the world. A greater danger than the surface fleet is the large force of nuclear-powered submarines, which can remain submerged for several months. In an effort to detect these elusive craft, the US Navy has developed a satellite-borne infra-red detector. This is intended to trace the minute temperature variations caused by the discharge of warm water used to cool the submarines' nuclear reactors. However it is believed that in practice this sensor has given disappointing results.

The Soviet Union uses radar-equipped reconnaissance satellites to detect and locate naval targets for attack by long-range antiship missiles. Targeting data is relayed to a surface ship which then launches its

The Soviet Union uses massive booster rockets, like the Vostok (above and below right) to put payloads into space. Its US equivalent is the Titan (right).

missile. These satellites are not placed in orbit on a regular basis, but have been used during Soviet naval exercises. One of these, Cosmos 954, achieved notoriety in January 1978, when it re-entered the atmosphere over Canada, contaminating a wide area with uranium from its nuclear reactor.

The vital task of communications increasingly depends on satellites, with a high proportion of long range messages being passed by this means. The United States' primary comsat system in the early 1980s is DSCS-2 (Defense Satellite Communications System, Phase 2), which carries teletype, data and voice communications for the three services. Six satellites are intended to give worldwide communications coverage, in contrast to the 26 satellites needed by the earlier DSCS network.

The DSCS-2 satellites are placed in a high geostationary orbit above the equator, with one above the Atlantic, another over the Indian Ocean, two stationed over the Pacific and two as back up systems. In the event of one satellite failing, another can be maneuvered by the ground control station to take its place. The next generation of comsats, DSCS-3, is currently under development. They will have an expanded capacity for message handling and will have a life of ten years – double that of DSCS-2.

The US Navy's communications needs are somewhat different from those of the other services and so it has its own comsat system. Named Fltsatcom, it has at least three satellites in stationary orbit at all times. This system provides priority communication for naval forces over most of the world, with more than 300 warships and 150 submarines equipped to use it. However, because these and the DSCS comsats are in orbit above the equator, the polar regions are poorly served.

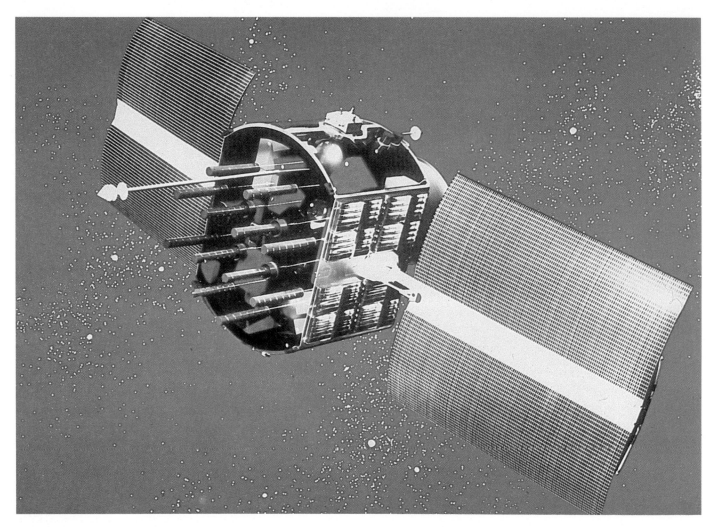

Accordingly the Satellite Data System (SDS) satellite was developed to meet SAC's need for high priority communications in the Arctic regions.

The Soviet Union uses two types of comsat. The Molniya system, which became operational in 1967, placed satellites in a high elliptical orbit, inclined at 65 degrees to the equator. These are now giving way to a series of comsats in geostationary orbit over the equator, a development which reflects the newly-developed global reach of the Soviet armed forces. A third class of Soviet comsat has been tentatively identified. These are placed in a low orbit and can relay signals from the low powered transmitters used in covert military operations.

The use of the stars for navigation is almost as old as seafaring, so it is fitting that the US Navy was the pioneer of the use of artificial stars, or navsats, for this important military requirement. During the 1960s the US Navy satellite program, named Transit, provided ships with accurate positional fixes by means of radio signals. The present navsat network, Navstar GPS (Global Positioning System), will become fully operational by 1987. It has world-wide coverage and is to be used by all three US services. It is entirely passive, so it can be employed without fear of betraying the user's position to the enemy. A ship's captain, a fighter pilot

The US Defense Satellite Communication System is the primary Department of Defense comsat network. The DSCS satellites (above) are in geostationary orbit. The Titan IIID booster (right) lifts such payloads as Big Bird reconsats into near-earth orbit.

or an infantryman can immediately learn his exact location at the push of a button.

The Navstar GPS will eventually comprise a total of 18 satellites positioned in three orbital rings inclined at 63 degrees to the equator. Continuously beamed signals will be picked up by a ground receiver, which will make use of those from four satellites to determine a position to within an accuracy of some 50 feet. The first Navstar was launched in 1978 and when the network is complete it is expected to have over 20,000 users every day. Its applications are manifold, ranging from accurate blind bombing to vectoring tactical fighters to a rendezvous with a tanker aircraft – all without the use of tell-tale radar or radio transmissions. Navstar also has strategic applications, including the guidance of missile warheads and the targeting of anti-satellite weapons. The equivalent Soviet navsats are much less complex and are mainly for use by warships.

The strategic early warning systems, which guard against a suprise attack by ICBM forces, have early

warning satellites as their first line of defense. These Program 647 satellites are placed in geostationary orbit around the equator. They carry infra-red detection equipment which can pick up the exhaust flare from a missile's rocket motors from ignition until shut down at the top of the trajectory. Consequently their use is limited to early warning rather than missile tracking. With early warning satellites over the Indian Ocean and Central America, the United States can detect launches from the Soviet ICBM fields and from SSBNs operating off the American coasts. The satellites also provide warning against the fractional orbit bombardment system (FOBS), tested but apparently not deployed by the Soviet Union. FOBS nuclear warheads sought to evade detection by the fixed BMEWS (Ballistic Missile Early Warning System) radars by following a partial orbit and descending over the United States' southern borders. A related military satellite mission is the detection of nuclear tests, by measuring radiation and electro-magnetic pulse.

Soviet early warning satellites cannot follow a geostationary orbit. In order to be positioned to cover the United States' ICBM fields and transmit warning back to the Soviet Union they must follow a highly elliptical, semi-geostationary orbit. The high points of the orbit are over the Northern Atlantic and Western Pacific, where the satellites appear to hover for a period of five to six hours. Nine must be in orbit to guarantee adequate coverage and back-up in the event of failure. However this goal is expensive, particularly as the Soviet early warning satellites suffer from quite a high failure rate, and has yet to be achieved.

As an inevitable result of the extensive military use of satellites, both Superpowers have developed anti-satellite (Asat) systems capable of knocking-out their opponent's vehicles in orbit. The Soviet Union has taken the lead with satellite interceptors, or killersats. At present these appear to be effective only against satellites in low orbits, but undoubtedly their range can be extended. The killersat is guided into an orbit which passes close to its target as soon after the killer's launching as possible, so that the victim cannot be maneuvered to safety. When the killersat reaches the closest position, a conventional warhead is exploded showering the victim with high-velocity fragments. The US has followed a quite different approach and is developing an anti-satellite missile, which will be launched from high-flying F-15 interceptors.

Many of the manned space missions have a military significance which tends to be overshadowed by the parallel scientific aims of these flights. A high proportion of the American astronauts and Soviet cosmonauts are military officers and they have undoubtedly carried out reconnaissance missions from space. Not only can a spacecraft's crew direct cameras and multi-spectral sensors to better effect than those in reconsats but they can also make visual observations and report directly back to earth. Additionally they can test equipment

which will later fly in unmanned vehicles.

The Soviet Union has developed the Salyut space station, which can remain in orbit for long periods manned by relays of cosmonauts. This program may lead to the construction of a large orbiting space station. The traditional method of using booster rockets to place payloads in space is being further developed by the Soviet Union. The Soviet 'super booster' is apparently made up from clusters of rocket motors, grouped into four stages, and is capable of placing a payload of almost 200 tons into low earth orbit. By contrast the United States' radical new space launch vehicle – the Shuttle – can lift only one-eighth of this load.

Direct comparisons between the Soviet 'super booster' and the Shuttle are misleading, however. This is because the Shuttle can be reused many times, whereas any booster is expended on one flight. Not only can the Shuttle place heavy loads in earth orbit, it can also carry a specially-designed Inertial Upper Stage,

Laser and particle beam weapons could revolutionize the military uses of space, but their development is fraught with difficulties. An airborne laser beam weapon has been tested by this NKC-135A aircraft.

which will carry payloads up to medium and high orbits. Initially the Shuttle will replace traditional boosters for satellite launches, but it also has the capability of building and then servicing a modular, manned spaced station. The United States has lagged behind the Soviet Union in such developments, but the military implications of orbiting space platforms have alarmed some highly-placed intelligence officers.

The fear is that the Soviet Union could mount a directed energy (laser or particle beam) weapon on an orbiting space station. This weapon could be used against satellites, but more significantly it could prove to be a highly-effective ABM system, able to destroy hundreds of missiles and warheads. Clearly if such a weapon was possible, it would have a highly destabilizing effect on the strategic balance. However the technical problems of producing space-based, directed energy weapons are enormous. Their power requirements are phenomenal and even if this difficulty is overcome, the weapon still has to be accurately directed

onto its target. It is unlikely that such technology will be perfected within the immediate future.

The stationing of any weapon of mass destruction in space has been specifically outlawed by treaty. However, such an agreement can be violated by any power seeking an advantage in strategic weaponry. Indeed the Soviet Union came close to such an infringement with its FOBS. Therefore it may be that the future will see an extension of the strategic nuclear battlefield into space. All strategic rockets enter space for a portion of their flight and ABM systems will seek to intercept them there. It is but a small extension of this activity to put warheads into orbit. In World War III the war in the air may easily expand beyond the earth's atmosphere and into space.

An A-10A Thunderbolt II of the UK-based 81st TFW tests his 'decelerons' before flying a sortie from one of the Wing's forward operating locations in West Germany. Decelerons are ailerons which also hinge upwards as shown to act as aerodynamic brakes.

3. TACTICAL COMBAT

Tactical Combat

The ability of aircraft to intervene in the land battle constitutes one of the most important roles in air warfare and a large proportion of the NATO and Warsaw Pact air forces are assigned to this task. Air forces by virtue of their speed and flexibility can help to blunt a Soviet blitzkrieg-style attack on NATO's Central Front while ground forces are still deploying to their defensive positions. The Soviet assault will likewise be accompanied by air attacks on NATO front-line forces and their rear areas. Both sides will attempt to control the air space over the battlefield to facilitate their own ground and air operations and to deny enemy aircraft the opportunity of influencing the land battle. Interdiction missions will be flown deep into enemy territory in an attempt to isolate the battlefield from reinforcement and resupply.

The operations of tactical air forces can be classified into three main areas of activity. Interdiction and tactical nuclear strike are carried out by long range attack aircraft with the ability to penetrate deep into the enemy's rear areas. The targets for interdiction aircraft will include transport systems, supply dumps and military formations moving to the battle area. Nuclear strike will be a last ditch defense, when conventional means have failed. Counter-air and air superiority missions will be directed against the enemy air forces, either on their own airfields or in the air. Similarly ground-based air defense systems including SAMs, AA artillery and their associated radars and support equipment will be attacked by specialized defense suppression aircraft, or alternatively they will be neutralized by ECM equipment. Finally the troops in contact with the enemy can call on direct close air support from ground-attack aircraft and they will be further assisted by battlefield interdiction sorties farther to the rear.

The USAF maintains a large tactical fighter force, which is assigned to Tactical Air Command (TAC) based in the United States, with additional reinforcement available from units of the Air National Guard (ANG) and Air Force Reserve (AFRES). Tactical fighter wings also serve with US Air Force in Europe (USAFE) and Pacific Air Forces (PACAF). These forces may be deployed anywhere in the world at short notice and consequently all tactical fighters are equipped for inflight refueling. Their supporting tanker aircraft are drawn from SAC assets, but TAC does control its own tactical reconnaissance, forward air control (FAC) and airborne warning and control aircraft and special operations units are also assigned.

According to the US Joint Chiefs of Staff military posture statement in 1982 the strength of the tactical fighter force was equivalent to 24 active wings and 12 reserve wings (in fact the strength on paper appears to be greater). The main aircraft types in service were the F-111 all-weather interdiction aircraft (240 on strength), F-15 Eagle air superiority fighter (396), F-16

A pair of F-15 Eagle air superiority fighters of TAC's 1st TFW (above) fly from their base at Langley, Va. The F-16 (inset) comes from the 388th TFW, Hill AFB, Utah.

Fighting Falcon multirole fighter (276), F-4 Phantom II multirole fighter (792), A-10 Thunderbolt II attack aircraft (444) and A-7 Corsair II attack aircraft (324).

The deployment of the tactical fighter force within TAC, PACAF and USAFE (non-fighter units are not listed) is as follows:

Tactical Air Command, HQ Langley AFB, Va.		
Unit	Base	Equipment
1st TFW	Langley AFB, Va	3 squadrons of F-15A
4th TFW	Seymour Johnson AFB, NC	3 squadrons of F-4E
23rd TFW	England AFB, La	3 squadrons of A-10A
27th TFW	Cannon AFB, NM	3 squadrons of F-111D
31st TFW	Homestead AFB, Fl	4 squadrons of F-4D
33rd TFW	Eglin AFB, Fl	2 squadrons of F-15A

8th TFW	Kunsan AB, Korea	2 squadrons of F-16A
18th TFW	Kadena AB, Okinawa	3 squadrons F-15C

US Air Forces in Europe, HQ Ramstein AB, West Germany		
Unit	**Base**	**Equipment**
20th TFW	RAF Upper Heyford, UK	3 squadrons of F-111E
32nd TFS	Soesterberg, Netherlands	1 squadron of F-15C
36th TFW	Bitburg, West Germany	3 squadrons of F-15C
48th TFW	RAF Lakenheath, UK	4 squadrons of F-111E
50th TFW	Hahn AB, West Germany	3 squadrons of F-16A
52nd TFW	Spangdahlem AB, West Germany	3 squadrons of F-4E/G
81st TFW	RAF Bentwaters and Woodbridge, UK	6 squadrons of A-10A
86th TFW	Ramstein AB, West Germany	2 squadrons of F-4E
401st TFW	Torrejon AB, Spain	3 squadrons of F-4D

AFRES and ANG units can be mobilized in an emergency to bolster the regular forces. Their effectiveness is being increased by supplying some units with factory-fresh A-10 and F-16 aircraft rather than the hand-me-downs which they have relied on in the past. Tactical air power in Europe will be augmented by some 3000 tactical aircraft of NATO Allied nations.

The Soviet equivalent of TAC is Frontal Aviation, the largest command within the Soviet air force with a front-line strength of some 4800 combat aircraft plus transports and helicopters. Soviet doctrine emphasises the co-ordination of all arms as the key to success in modern warfare and this is reflected in close links between Frontal Aviation and the ground forces it supports. Frontal Aviation is composed of 16 air armies, which are attached to the 12 military districts of the USSR and to the Soviet forces in Czechoslovakia, East Germany, Hungary and Poland. The air armies are made up of divisions, each of which fulfills a specific role such as close air support. The regiment is the basic operational unit and each flies a single aircraft type. There are usually three regiments to a division and the regiment is made up of three squadrons each with some 18 aircraft. Air armies can vary widely in strength, with the 16th Army attached to the Group of Soviet Forces in Germany fielding 1000 aircraft in contrast to the 5th Army's strength of only 100 in the Kiev Military District. The heaviest concentration is of course on the central sector of the European theater.

The 1970s saw the transformation of Frontal Avia-

35th TFW	George AFB, Ca	3 squadrons of F-4E
37th TFW	George AFB, Ca	3 squadrons of F-4G
49th TFW	Holloman AFB, NM	3 squadrons of F-15A
56th TFW	MacDill AFB, Fl	4 squadrons of F-16A
347th TFW	Moody AFB, Ga	3 squadrons of F-4E
354th TFW	Myrtle Beach AFB, SC	3 squadrons of A-10A
366th TFW	Mountain Home AFB, Id	3 squadrons of F-111A
388th TFW	Hill AFB, Utah	4 squadrons of F-16A
474th TFW	Nellis AFB, Nev.	3 squadrons of F-16A

Pacific Air Forces, HQ Yokota AB, Japan		
Unit	**Base**	**Equipment**
3rd TFW	Clark AB, Philippines	2 squadrons of F-4E/G

tion from a force equipped for defensive counter-air missions, with a secondary close support role, to a more balanced tactical air force with greatly increased offensive capabilities. The numerical and qualitative improvements to Frontal Aviation's forces, together with an increasing emphasis on offensive missions against NATO ground and air forces, represent an enhancement of the Soviet Union's military capability which is every bit as significant as the far more publicized expansion of the Soviet Navy.

The key to Frontal Aviation's greatly expanded capabilities has been provided by two significant new combat aircraft – the MiG-23/27 Flogger family and the Su-24 Fencer – and also by modification to older designs. There are some 1400 Floggers operational with Frontal Aviation and these include both Flogger-B and G air superiority fighters and Flogger-D and F ground attack fighters. Almost as numerous are late-model MiG-21 Fishbeds (1300) which are all-weather air superiority fighters, with secondary ground attack capabilities. Since its service introduction in 1959 as a clear-weather, short-range point-defense fighter, the MiG-21 has been continually upgraded to improve its all-weather and payload/range capabilities. Another example of the evolutionary process is provided by the Su-17 Fitter-C, D and H (650 operational). These are

variable-geometry modifications of the earlier fixed, swept-wing Su-7 Fitter-A (200 remain in service). Finally there are some 400 of the formidable (by Soviet standards) Su-24 Fencer interdiction aircraft, which can cover targets throughout most of Western Europe.

In addition to its tactical fighter aircraft, Frontal Aviation controls a large force of battlefield helicopters and has its own reconnaissance, transport and ECM support aircraft. Nearly 70 percent of Frontal Aviation's units are deployed in Europe or the western USSR, where there are a large number of airfields to accommodate them. This is in contrast to the NATO powers, which face an acute shortage of military airfields (for example RAF Germany operates from only four). The Soviet Union is supported (but with what degree of enthusiasm is open to doubt) by her Warsaw Pact allies. They can add some 2400 tactical aircraft to the Soviet strength, but many are elderly and their greatest contribution is likely to be the air defense of rear areas against NATO interdiction aircraft.

Isolation of the battlefield from reinforcement and resupply by interdiction of enemy lines of communication is an attractive aim. In practice it has often proved very difficult to accomplish, as the USAF campaign against the Ho Chi Minh Trail during the Vietnam War bears witness. However in a European war transport

Representative of the new generation of Soviet tactical fighters are the Su-17 Fitter-C (above left) and the MiG-23 Floggers (above).

targets will be more vulnerable and even a comparatively short delay in feeding reinforcements into the battle area may prove to be crucial. The Soviet army is especially vulnerable in this respect, as its divisions have less staying power than their NATO equivalents and Soviet doctrine requires that they be replaced regularly by fresh divisions from the rear. Therefore any disruption of roads, railways, bridges or staging areas may result in the attack losing momentum.

However NATO forces are far from being immune to the effects of interdiction. They too rely on reinforcement to conduct a successful defense and some units will have to come from the United States. Much of the heavy equipment of these divisions is stored in Federal Germany and represent as attractive a target as the forces themselves. Nuclear weapons storage sites are also likely to be a priority target and headquarters and command and control centers will also be attacked.

Interdictor aircraft will be employed on offensive counter-air missions against enemy airfields. They can also directly attack reinforcements moving up to the battle area. These troops are likely to be less prepared to meet air attack than those already deployed. SAM and AA fire will be less intense and forces on the move will offer a better target than those dispersed over the battlefield.

Although NATO's strategy is one of 'flexible response', so that a conventional (i.e. non-nuclear) attack will be met at first with conventional defenses, it is becoming increasingly recognized that NATO's numerical inferiority may force the alliance into the use of tactical nuclear weapons to stem a Soviet breakthrough. Whether the Soviet marshals would countenance the first use of nuclear weapons to precipitate the collapse of NATO forces on the central front is open to doubt. They certainly appear to show little reluctance in contemplating fighting in a nuclear, bacteriological or chemical (NBC) warfare environment. Whatever the imponderables about the use of such weapons, it is certain that many NATO and Warsaw Pact aircraft can use them. They include the Soviet Su-24 Fencer, Su-17 Fitter and MiG-27 Flogger. NATO nuclear-capable aircraft are the F-111, F-4, Jaguar, Tornado and F-104.

The characteristics generally demanded of an interdiction/strike aircraft are good range and payload, the ability to evade air defenses by flying at low level and following the contours of the terrain and the capability of operating in all weather conditions. The pre-eminent

example of such a warplane in the USAF's armory is the General Dynamics F-111. The F-111 was designed to meet a controversial requirement, calling for a multi-role aircraft to undertake close air support and interdiction for the USAF combined with fleet air defense for the US Navy. In the event only the interdiction version entered service, together with the FB-111 strategic bomber.

The F-111 is a two-seat, variable-geometry wing fighter-bomber, powered by two Pratt & Whitney TF-30 afterburning turbofans. The variable-sweep wing enables the F-111 to operate from relatively short runways, yet it can reach supersonic speed at low altitude and Mach 2.5 above 60,000ft. In the fully-developed F-111F version the powerplants are TF-30-P100 turbofans each developing 25,100lbs of thrust with afterburning. Takeoff weight is 85,000lbs and the aircraft can carry a 6000lb bomb load over a combat radius of 1000 nautical miles. Alternatively a maximum warload of 24,000lbs can be lifted to a radius of 350

nautical miles. All tactical versions of the F-111 span 63ft (unswept), with a length of 73ft 5in and a 20mm M-61A Vulcan rotary cannon is standard armament.

There are four tactical versions of the F-111 in service with the USAF, plus the strategic FB-111. The F-111A was the initial production version for TAC and 86 remain in service. This version forms the basis of the EF-111A ECM aircraft, 42 of which are to be rebuilt from F-111A airframes. The B model was to have been the US Navy's fighter and the C is a variant of the A supplied to the Royal Australian Air Force. The F-111D, E and F all incorporated progressive improvements to the powerplants and navigation and attack avionics. Further upgrading is in prospect (including improvements to offensive and defensive avionics and provision of more advanced weapons) to keep the F-111 fully effective into the 1990s.

Even when the planned strike versions of the F-16 or F-15 enter service, it is estimated that 30 percent of interdiction targets in Eastern Europe can be reached

Four of the European NATO allies fly the F-16A, including the Royal Netherlands Air Force.

a picture of the target area by day or night. He can then acquire his target and release weapons against it, further assisted by the Pave Tack pod's laser range-finder and designator. The USAFE's 48th TFW flying F-111Fs at Lakenheath, UK, is the first user of Pave Tack, but the system may later be fitted to F-111Ds and Es.

The USAF's second interdictor strike aircraft is the ubiquitous F-4 Phantom multi-role fighter. Although a less capable aircraft in the specialized interdiction role (combat radius with 6000lbs of bombs is 350 nautical miles, compared with the F-111's 1000nm radius with the same load), the F-4 is available in greater numbers. However this classic fighter is now past its prime and is being replaced by the General Dynamics F-16 in the tactical fighter wings. A specialized and much modified strike version, the F-16E, is competing with the F-15E to meet the USAF's requirement for a new attack aircraft. Another possible contender is the highly-secret Lockheed CSIRS (covert survivable in-weather reconnaissance/strike) fighter, which uses Stealth technology. However this is at present a high-risk research program, with only 20 aircraft in prospect. The US Navy can also take part in an interdiction campaign with its carrier-based Grumman A-6E Intruders.

America's NATO allies contribute to the Alliance's interdiction/strike capabilities, notably with the Panavia Tornado, which in 1982 began to enter service with the air arms of the UK, Germany and Italy. In the RAF, Tornado GR Mk 1s will replace the six squadrons of Vulcan B Mk 2 bombers scheduled to disband in 1982 and will supplement the Buccaneer S Mk 2 force. The Buccaneer was originally a shipboard strike/attack aircraft which was pressed into RAF service to fill a gap created by the cancellation of the TSR-2 and F-111K programs. It has proved to be extremely successful, although one unforeseen setback was the grounding of the entire force in 1980 due to fatigue problems. This has resulted in the numbers in service being reduced. Despite its age (first flight 1958) the Buccaneer is a stable, maneuverable and relatively fast (Mach 0.9) weapons platform for interdiction/strike at low level and its payload/range characteristics are impressive. Three squadrons are assigned to that role (No 208 at Honington, UK and Nos 15 and 16 in Germany), although the two squadrons in Germany will eventually re-equip with the Tornado and the remaining squadron will probably convert to the maritime strike role. Buccaneers have recently been given a measure of self-defense capability by the fitting of Westinghouse AN/ALQ-101 ECM jamming pods and Sidewinder AAMs. Their weapons load of 16,000lbs can include Paveway laser guided bombs.

Yet it is the Tornado which will give the NATO European allies a significant ability to carry out long range interdiction/strike. With the RAF it will upgrade an existing force, but for West Germany and Italy it represents a new capability in comparison with the

by the F-111 alone. The F-111's ability to operate in all weathers is also valuable. In a NATO exercise in Europe F-111s were scheduled to undertake 30 percent of all USAF sorties, but in the event bad weather intervened and the F-111s contributed 86 percent. However on the debit side, if the weather cannot ground the F-111 force, spares shortages and unscheduled maintenance requirements often can. Such problems have dogged the aircraft throughout its career.

The process of improving the F-111's target acquisition and weapons capability has already begun, with the introduction of the Pave Tack weapons aiming system which is used in conjunction with such precision-guided munitions (PGM) as the GBU-15 glide bomb and Maverick TV-guided ASM. Pave Tack is a pod containing a forward-looking infra-red (FLIR) system, which presents the weapons system officer with

The F-111 fighter-bomber provides the USAF with a formidable all-weather attack capability (below). An F-111D of the 27th TFW (Left) releases Mk 82 practice bombs. A SEPECAT Jaguar tactical fighter of the Armée de l'Air maneuvers at low level (below left).

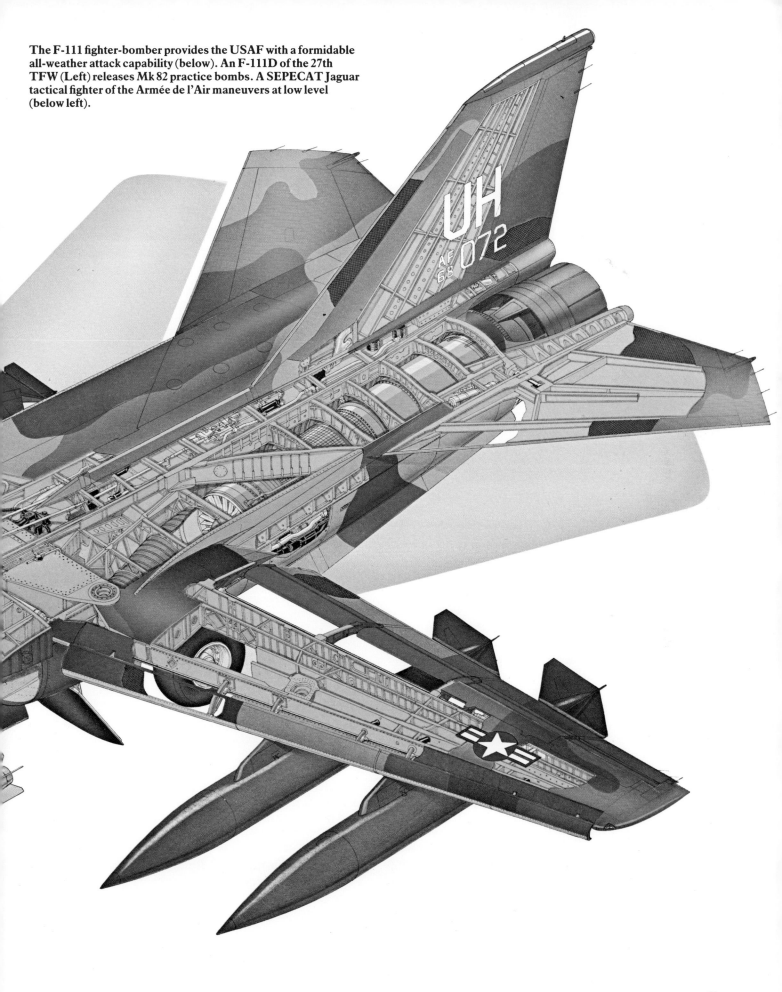

The US GBU-15 electro-optically guided glide bomb (above) makes possible precision attack from stand-off ranges. The RAF's Tornado GR Mk 1 (above right) is specially designed for low-level interdiction. The USAF's F-4 Phantom (right) has proved to be a versatile tactical fighter.

Lockheed F-104 Starfighter which it will replace. The UK requirement is for 220 Tornado GR Mk 1 interdictor/strike aircraft, with a further 165 Tornado F Mk 2 air defense interceptors. West Germany needs 212 aircraft for the Luftwaffe and 112 for the Navy's maritime attack squadrons, while Italy plans to acquire a total of 100. By April 1982 100 had been delivered, with the manufacturing workload being divided amongst the three participating nations. Conversion training is also a co-operative effort and the Trinational Tornado Training Establishment has been set up at Cottesmore in the UK.

Tornado is a compact, variable-geometry aircraft, which accommodates a crew of two seated in tandem. Power is provided by two efficient and economical Turbo-Union RB.199 turbofans, each giving 16,000lbs thrust with afterburning. Wingspan (in the fully forward position) is 45ft 7in and length 54ft 9in, with maximum loaded weight 58,400lbs. Maximum speed ranges from Mach 1.1 at low level to Mach 2.2 above 60,000ft. Tactical radius flying a lo-lo-lo profile is 450 miles and external weapons load is 16,000lbs. A built-in armament of two 27mm Mauser cannon is standard. The landing run is impressively short at some 500ft. Tornado therefore combines performance and agility with the range/payload characteristics usually associated with far more staid aircraft.

A very comprehensive and sophisticated avionics fit assists the Tornado crew. The heart of the system is a central computer, which can store navigational and mission data and display it on request. Navigational aids include an inertial system and doppler radar. A terrain-following radar allows the aircraft to fly at high speed and low level automatically maintaining a predetermined height (typically 200ft) above the ground. There is also a ground mapping attack radar, which provides pilot and navigator with computer processed information. The final stages of the attack are facilitated by a nose-mounted laser rangefinder. Weapons release can be automatically controlled by the central computer. Finally much of the turbulence and buffeting associated with low level flying is damped out by the Tornado's Command and Stability Augmentation System.

A low flying Tornado will present a difficult target to enemy air defenses, and their problems will be compounded by the aircraft's ECM equipment, which includes a radar warning receiver, Sky Shadow jam-

The Soviet Union's Sukhoi Su-24 Fencer (above) represents a dramatic increase in power for Frontal Aviation, as it is the first Soviet tactical warplane capable of far-ranging interdiction missions. The A-7K Corsair II (right) is intended to train US Air National Guard units on the single seat A-7, one of the capable new warplanes for the reserves.

ming pods and chaff and flare dispensers. Tornado also has teeth and an unwary Soviet interceptor could find himself on the receiving end of its 27mm cannon or AIM-9L Sidewinder missiles. A wide range of offensive armament may also be carried, embracing free fall conventional and nuclear bombs, PGMs, anti-armor and anti-radar missiles, the RAF's JP233 anti-airfield cluster bomblet dispenser and the Luftwaffe's similar MW-1 weapon.

Frontal Aviation's interdiction/strike fighter, the Sukhoi Su-24 Fencer, is in many ways a smaller and lighter version of the F-111, although it is less technologically advanced. It is the first Soviet tactical fighter to carry a second crewmember and this reflects an increased emphasis on all-weather capability and weapons' management over such ground attack aircraft as the MiG-27. The Su-24 is the least known of the present generation of Soviet warplanes and perhaps for this reason it has become as much of a 'bogey' in the minds of Western defense planners as has Backfire.

Fencer has a variable-geometry wing, with a sweep-back angle of 16 degrees in the full-forward position, increasing to 68 degrees when fully swept. Span with wings in the forward position is 56ft 10in and length 72ft 10in, with takeoff weight being in the region of 65,000lbs. Power is provided by two afterburning turbofans, which produce an estimated 19,500lbs of thrust each. Performance includes a maximum speed of Mach 2 at 36,000ft reducing to Mach 1.2 at sea level, with combat radius in the order of 700 miles (assuming a hi-lo-hi profile). The two crew members are seated side-by-side in a rather cramped forward cockpit, with an attack and probably a terrain-following radar in the nose. A feature of the design is the rugged undercarriage, optimized for operations from short, rough forward airfields. The built-in armament consists of a twin-barrel GSh-23 23mm cannon and there are seven external hardpoints, but no internal bay. Weapons load is some 12,500lbs and may be made up of free-fall bombs or ASMs, including AS-7, AS-9, AS-11 or AS-12. AA-2 Atoll and AA-8 Aphid AAMs can be carried.

By making use of economical hi-lo-hi flight profiles, operating from forward bases and reducing its weapons load to 4000lbs, Fencer is able to operate over most of Continental Europe and much of the UK, particularly the important concentration of USAF and RAF air bases in East Anglia. Yet it is the classic interdiction targets which Fencer is most likely to seek out and thus fill the gap between the MiG-27 Flogger operating over the battlefield and the long-ranging Backfire.

The successful accomplishment of interdiction, counter-air, defense suppression or close-air support missions depends as much on the weapons employed as on the aircraft carrying them. There is a bewildering array of such weapons in service, both conventional and nuclear, although there is little information on the nuclear systems. The free-fall bomb is still in widespread use and the US Mk 80 series of low-drag, general purpose, high explosive bombs is in many ways typical. They are available in four main classes: the Mk 81 of 250lbs, the Mk 82 of 500lbs, the Mk 83 of 1000lbs and the Mk 84 of 2000lbs. The Mk 81, 82, and 83 can be fitted with a retarding tail device (when they are known as Snakeye bombs) which enables them to be released at low level without ricocheting off target or damaging the releasing aircraft in the blast. Fuzing can also be varied according to the mission, the options including proximity, for an airburst over the target, instantaneous on impact, or short and long delays.

Alternatives to high-explosive bombs include fire-bombs and cluster bomb units (CBUs). The latter are dispensers filled with a large number of small munitions, which on release are scattered over a wide area. They are thus very effective against dispersed targets. CBUs can be loaded with antitank, antipersonnel or antimateriel munitions, according to the anticipated target. This idea has been extended to large dispenser pods attached to the aircraft which eject submunitions. The RAF's JP233 anti-airfield weapon is such a system, dispensing both concrete-penetrating submunitions and antipersonnel devices to hamper repair work. West Germany's MW-1 is similar, but it is a multi-purpose weapon and can be most effective against enemy tank concentrations.

Unguided rockets, because they can be fired in salvos, have a high probability of hitting a target on the first firing pass. They are usually carried in pods, in the case of the US 2.75in FFAR (folding fin aircraft rocket) holding seven or 19 rockets. Warheads can be anti-tank, fragmentation, smoke (for target marking) or white phosphorous.

Despite the number and variety of unguided weapons in current use, it is the precision-guided munition that has made the greatest impact on modern air warfare. Laser and electro-optically guided bombs have made possible a hitherto unprecedented degree of precision in tactical bombing. This is perhaps best illustrated by an episode in the Vietnam War. The Thanh Hoa bridge had defied the best efforts of US strike aircraft to destroy it since 1965, but when laser-guided bombs were used against it in 1972 the bridge was severely damaged during the first attack. The Paveway family of laser-guided bombs and Walleye TV-guided glide bomb are both in US service. They are to be followed by the GBU-15 glide bomb, which may have TV or imaging infra-red guidance which is not dependent on good light conditions.

The 'stand-off' capability of glide bombs is particularly useful against heavily defended targets. Significantly the USAF pulled out of the JP233 program (which was to have been a joint UK/US venture) because it disliked the attacking aircraft having to overfly its target. The answer to this problem is the air-to-surface missile, which has a better stand-off capability than the glide bomb, but it is also far more expensive. The widely-used Hughes AGM-65 Maverick has a range of some 14 miles and can use TV, laser or imagining infra-red guidance. An altogether more specialized ASM is the anti-radiation missile, such as the Shrike or Standard ARM, which homes onto radar emissions. It is particularly useful for defense suppression and an advanced new missile, the AGM-88 HARM (high-speed ARM) is being developed. Other interesting new lines of development include the Wasp millimeter-wave radar guided mini-missile for anti-armor use, which can be carried in 12 round pods and the laser-guided Hypervelocity missile, which relies on kinetic energy rather than a conventional warhead to knock-out the enemy tank.

Little is known about Soviet developments in PGMs although it is reported that Frontal Aviation has laser-guided bombs in service. Tactical ASMs include the radio-command guided AS-7 Kerry, with a range of six miles. Soviet aircraft are also armed with bombs and unguided rockets and Frontal Aviation almost certainly also uses chemical and bacteriological weapons (the US holds stocks of nerve-gas bombs as a deterrent). NATO air forces frequently practice operating in a nuclear or bacteriologically contaminated environment. This involves air and ground crews wearing restricting and uncomfortable NBC protective clothing and carrying out decontamination procedures on aircraft and equipment.

However destructive a weapon may be, it will be of little use unless it is delivered accurately and therefore navigational and aiming systems are of great importance in attack aircraft. All-weather operations are particularly demanding in this respect. Radar provides much of the information needed, including ground mapping, terrain avoidance, height above terrain and the acquisition, tracking and ranging of ground targets. However radar emissions can betray the presence of the attacking aircraft and they may also be jammed. So various alternative navigation/attack systems are under development, including millimetric radar, which is harder to detect and jam than the current radars operating on centimetric wavelengths.

Infra-red imaging equipment can provide a TV-type picture of the landscape in virtually all weathers, although cloud and rain will degrade its performance. The US LANTIRN (low altitude navigation and

targeting infra-red system for night) is being developed to provide single-seat F-16 and A-10 aircraft with all-weather attack capability. It combines FLIR with a terrain-following radar and laser target designation. Low-light level television (LLTV) can perform the same function as FLIR on a clear night, but it would be even more affected by adverse weather. Inertial navigation systems can direct an aircraft to a predetermined target with no external source of reference. The aircraft's starting point is precisely determined and fed into a computer, all subsequent movement is recorded by sensors and information on the aircraft's position and directions to the target or to a 'waypoint' are displayed to the crew. The route can follow a series of dog-legs to avoid heavily-defended areas. The system can suffer from 'drift', but accuracy is typically within one nautical mile for every hour of flight time. Other navigational aids which could be of help to the attack

aircraft are the Navstar GPS satellites and the TERCOM system developed for cruise missiles.

Weapons aiming systems rely on a computer to calculate the impact point for bombs, rockets or gunfire, taking account of the velocity of the attacking aircraft, the slant range to the target, the aircraft's attitude in flight and the ballistic characteristics of the weapons. Most modern attack aircraft make use of laser rangefinders and targets can also be marked by laser for attack by laser-guided weapons. Laser designation can be used by ground controllers to indicate a target for attack by PGMs, unguided or 'dumb' bombs, rocket or cannon fire during close air support sorties. Target indication, weapons release instructions and steering commands are projected onto a head-up display (HUD), which is basically similar to the traditional gunsight, but provides the pilot with navigational and flight instrument data as well as weapons release

The West German Luftwaffe plans to equip its Tornados with the MW-1 submunition dispenser (above). Weapons aiming is assisted on modern warplanes by the head-up display (inset) which projects flight data and weapon aiming commands into the pilot's field of view.

instructions. Thus the pilot can complete his attack without having to look down into the cockpit for essential information.

The air superiority – or counter-air mission, as it is often called – is crucial to the successful conduct of tactical air warfare. Air superiority forces seek to gain a sufficient degree of domination over their opponents to allow friendly air and ground operations to be carried on without undue interference. The virtual destruction of the enemy's aircraft and weapons systems is an ideal that is seldom achieved – the 1967 Arab-Israeli War being a rare instance, accomplished by an Israeli pre-emptive attack in the opening hours of the conflict. In practice, however, air superiority operations will be much more localized both in time and space, for example seeking to provide cover for a newly-established bridgehead during a river crossing, or clearing the path for a force of attack aircraft. A certain amount of enemy air activity must perforce be tolerated, but it cannot be allowed to deflect air or ground forces from their objectives.

Air superiority missions can be classified in various ways. There are offensive counter-air operations which seek out the enemy's air force over his own territory and on his airfields, while defensive counter-air units attack enemy aircraft attempting to penetrate friendly airspace. Active counter-air operations aim to destroy the enemy's aircraft, missiles and their support installations (e.g. airfields), while passive measures are designed to limit the effects of air attack by the use of cover, dispersal, deception, shielding and damage control. However, perhaps the most satisfactory breakdown of the air superiority mission is by the tactics employed: these are airfield attack, air-to-air fighter combat and defense suppression (the attack of ground-based systems, such as missiles and their radars).

The air superiority fighter may be called on to escort a force of attack aircraft or to patrol over a sector of the battlefield and its rear areas to keep the skies clear of enemy aircraft. It therefore requires good range or endurance. As it is required to engage high-performance enemy aircraft in air-to-air combat, it must be agile and fast. Finally as its primary role is air combat it must be heavily armed with AAMs and cannon. Not all air superiority fighters follow this design philosophy, which is particularly hard to fulfil satisfactorily because the demands of endurance and performance are to a large extent conflicting. An entirely different view of the air superiority fighter is exemplified by such aircraft as the Soviet MiG-21 and US Northrop F-5 (which significantly is not in front-line service with the USAF). This doctrine calls for large numbers of lightweight, high performance aircraft in the belief that weight of numbers will tell over a smaller force of individually more capable aircraft. The Soviet Union has always favored numbers over quality, a position it has been forced into by the US pre-eminence in technology.

Air superiority is numbered amongst the roles of the versatile McDonnell Douglas F-4 Phantom, although this aircraft is now giving way to the more modern F-15 and F-16. The USAF flies three fighter versions of the Phantom, the F-4C, F-4D, and F-4E (plus the specialized F-4G for defense suppression and RF-4C tactical reconnaissance aircraft). The F-4 was originally developed as a two-seat shipboard fighter for the US Navy and such was its outstanding performance that it was adopted by the USAF. The first F-4Cs became operational in 1964 and for more than a decade the F-4 was the backbone of the tactical fighter force. Its achievements during the war in Southeast Asia ranged from air-to-air combat, through tactical bombing to defense suppression.

The initial USAF Phantom, the F-4C, was similar to the US Navy's F-4B with added air-to-ground attack capability. It was followed by the F-4D, which had an improved bombing computer and gunsight. The definitive F-4E introduced built-in gun armament (a 20mm M-61 Vulcan cannon) thus correcting a major shortcoming and the wing was fitted with leading edge slats to improve maneuverability. The F-4E spans 38ft 5in, is 63ft long and maximum takeoff weight is 59,000lbs. Power is provided by two 17,900lbs thrust J79-GE-17 turbojets with afterburning, giving a maximum speed of Mach 2.4. Service ceiling is 70,000ft and combat radius 900 miles. Air-to-air armament in addition to the M-61 cannon, comprises four AIM-7 Sparrow radar-guided AAMs and four infra-red AIM-9 Sidewinders. A wide range of bombs, rockets and ASMs can also be carried. Five of the NATO allies fly the F-4 (Spain, Greece, Turkey, West Germany, and Britain).

Unlike the F-4, the McDonnell Douglas F-15 Eagle was tailored to the air superiority mission from its inception. The F-15 became operational in 1975 and export customers include Israel, Saudi Arabia and Japan. It is a large aircraft for a single-seat fighter, spanning 42ft 10in with a length of 63ft 9in and maximum takeoff weight of 68,000lbs. The keys to the F-15's impressive performance are its two Pratt & Whitney F100-PW-100 turbofans each developing 23,900lbs of thrust with afterburning. Maximum speed is over Mach 2.5 above 36,000ft, yet fuel consumption is not excessive. When fitted with FAST (fuel and sensor tactical) pack conformal fuel pallets the ferry range is more than 3000 nautical miles. The large wing (608 sq ft) makes it highly maneuverable because wing loading is very low.

At first sight the F-15's armament appears to be no more effective than that of the F-4E, comprising four

F-16A tactical fighters of the 56th TFW fly in echelon formation (above right). The F-15 Eagles (right) are much heavier and more powerful, although both fighters undertake the air superiority role. The A-10A attack aircraft can locate targets marked by laser, using the Pave Penny tracker (far right).

70

Sparrow and four Sidewinder AAMs and a built-in M-61 20mm rotary cannon. However both missiles have been progressively improved since their service introduction and replacements are now under development. A new 25mm rotary cannon was in prospect, but development problems led to its cancellation. Nevertheless the F-15 does offer a substantial improvement over the F-4E by virtue of its APG-63 pulse-doppler radar, which enables it to detect and engage low-flying targets with a greater chance of success. Another improvement, compensating for the loss of a weapons system officer, is the positioning of weapons and radar switches on the control column and throttle, which allows the pilot to fight his engagement without needing to take his hands from the controls. Similarly a HUD gives him essential information without need to look down at his instrument panels.

Essentially a complementary design to the F-15, the F-16 Fighting Falcon was developed in response to the USAF's Lightweight Fighter program. The US requirement is for 1388 F-16s and the fighter is also in production in Europe for the air forces of Belgium, Denmark, the Netherlands and Norway. It has been blooded in action by the Israeli air force, which used it in the successful raid on Iraq's nuclear reactor in June 1981.

The F-16's powerplant is a single Pratt & Whitney F-100-PW-200 turbofan of 23,840lbs thrust with afterburning, essentially the same engine as that fitted to the F-15. Maximum speed is over Mach 2 at 40,000ft, reducing to Mach 1.3 at sea level. The F-16 is a small aircraft in comparison with the F-4, which it is gradually replacing in the USAF's Tactical Fighter Wings. Wingspan is 31ft and length 49ft 6in, with maximum takeoff weight at 33,000lbs. As with the F-4, the F-16 has dual air-to-air and ground attack capability. Its weapons load includes a built-in M-61 cannon, wingtip mounted Sidewinder AAMs (a second pair may be

The MiG-21 (top left) remains an important fighter with the Soviet air force and the Warsaw Pact allies. Many F-4 Phantoms continue to serve in NATO (top); a USAF F-4D leads a Luftwaffe F-4F (background) and RAF Buccaneer. The RF-4C (above) is used for tactical reconnaissance, this aircraft serving with the USAF's 363rd TRW. F-15 Eagles (left and above center) were designed from the outset for air-to-air combat and the view from the cockpit is particularly good.

carried under the wing) and up to 12,000lbs of bombs. Tactical radius with a 3000lb bomb load is 340 miles.

A compact radar, the Westinghouse AN/APG-66, is mounted in the aircraft's nose. It is a multi-mode sensor, which can be used both in air-to-air and air-to-ground combat. Radar controls are stick and throttle-mounted and air-to-air radar information can be projected onto the pilot's HUD. The F-16's cockpit incorporates a reclining seat, angled 30 degrees to the rear, which increases the pilot's 'G' tolerance over conventionally mounted ejection seats. Another innovation is the side-stick control mounted on the right of the cockpit, which replaces the centrally-positioned control column. Compared to the F-4 (which is twice the weight), the F-16 has thrice the combat radius for the air superiority mission. It is highly maneuverable, acceleration and turn rates being better than the F-4's by 60 percent.

Various modifications are under development as part of the Multinational Staged Improvement Program (they will be incorporated into European-built and USAF F-16s). The changes include an improved APG-66 radar and HUD, increased computer capacity and new cockpit displays. Improved weapons in prospect are Maverick ASMs, the GEPOD 30mm pod-mounted cannon for ground-attack and eventually the AMRAAM (advanced medium range air-to-air missile). ECM will be improved by the new Advanced Self-Protection Jamming system (ASPJ).

The F-16 is really only a lightweight fighter in com-

The West German navy's F-104 Starfighters are to be replaced with Tornados; a TF-104G is shown (above). Another two-seat conversion trainer is the Danish F-16B pictured (above left). An unusual feature of the F-16's cockpit (left) is the sidestick control in place of the conventional control column.

parison with its stablemate the F-15. The Lockheed F-104 Starfighter which the F-16 is replacing with many NATO air arms, more truly justifies this description. Originally intended as a clear weather interceptor, the Starfighter achieved an outstanding flight performance at the expense of payload and all-weather capability. The multi-role F-104G has undertaken the air superiority, interceptor, reconnaissance and ground attack roles with ten NATO air forces: Belgium, Canada, Denmark, West Germany, Greece, Italy, the Netherlands, Norway, Spain and Turkey. The F-104G spans 21ft 11in, is 54ft 9in long and maximum weight is 28,780lbs. It is powered by a single General Electric J79-GE-11A turbojet of 15,800lbs thrust with afterburning. Top speed is Mach 2.35 at 36,000ft and combat radius 745 miles. Built-in armament is a 20mm M61 Vulcan rotary cannon and weapons load is up to 4000lbs.

The Northrop F-5 series of lightweight fighters also serves with several NATO air forces and numerous Third World nations. The F-5A model is a supersonic (Mach 1.4) clear-weather fighter armed with wingtip-mounted AIM-9 Sidewinders, or a 6000lbs warload for ground attack. Power is supplied by two General Electric J85-GE-13 turbojets, each producing 4080lbs of thrust with afterburning. Maximum takeoff weight is 19,860lbs and dimensions are 25ft 3in span and 47ft 2in length. Versions of the F-5A serve with the air forces of Canada, Greece, the Netherlands, Norway, Spain and Turkey. The improved F-5E (with more powerful engines, increased fuel load and an APQ-153 radar) serves with the USAF and US Navy in the air combat training role. A further improvement the F-5G, powered by a single General Electric F404 turbofan (16,400lbs thrust with afterburning) has yet to fly, but a top speed of Mach 2 is predicted.

France's highly successful Mirage III series of fighters are multi-role aircraft which have been widely exported, but apart from those with the Armée de l'Air the only NATO Mirage IIIs are those serving with the Spanish and Belgian air forces. A tail-less, delta-wing aircraft, the Mirage III spans 27ft and is 49ft 3in in length. The IIIE version is powered by a 13,670lbs thrust SNECMA Atar 09C afterburning turbojet and has a maximum takeoff weight of 29,760lbs. Top speed is Mach 2.1 at 40,000ft and combat radius is 745 miles. Armament comprises two DEFA 30mm cannon, plus three AAMs. A simplified ground-attack version, the Mirage 5, equips one French escadre (wing) and another flies the Mirage IIIR tactical reconnaissance aircraft. From 1983 the Mirage III will be replaced by the Mirage 2000. This is a Mach 2.3 tail-less delta, which is to be built in three versions for the Armée de

The MiG-23 (cutaway artwork) is an impressive Soviet multi-role fighter.

l'Air: a multi-role fighter, an interceptor and a two-seat interdictor-strike aircraft.

For many years the backbone of Frontal Aviation's tactical fighter force was the MiG-21 Fishbed. This fighter is now being replaced by the MiG-23 Flogger, but nevertheless many MiG-21s remain in service. The first production version, the MiG-21F Fishbed-C, entered service in late 1959. It was a true lightweight air superiority fighter with no all-weather capability, a short endurance and relatively light armament. Over the succeeding 20 years there have been numerous improvements to the basic design, aimed at improving its all-weather performance and suiting it to the tactical fighter-bomber role. A basic fault that has never been satisfactorily eradicated is the MiG-21's short endurance.

The aircraft has many positive virtues to offset its fundamental weakness. It is highly maneuverable for a fighter of its period, although in comparison with the F-16 it is inferior in rate and radius of turn and in acceleration. It is easy to fly and maintenance, refueling and rearming present few problems. The gun armament of the late-model MiG-21bis comprises a twin-barrel GSh-23 23mm cannon, which is reliable and effective. Missile armament, usually four infra-red AA-2 Atolls, is less satisfactory because these weapons are not maneuverable enough for modern combat. However the later AA-8 Aphid is a much better weapon.

The MiG-21's small size gives its pilot an advantage in combat because it is more difficult to pick up than larger Western fighters such as the F-15 or F-4. Details of the MiG-21bis (Fishbed-N) include a span of 23ft 6in, length 51ft 9in and maximum loaded weight of 22,000lbs. The powerplant is a 16,500lb Tumansky R-25 afterburning turbojet, giving a maximum speed of Mach 2 at 36,000ft. Service ceiling is about 50,000ft and combat radius 300 miles. Rocket pods or AS-7 Kerry ASMs can be carried for ground attack, but

This MiG-23 Flogger (above) carries IR AA-8 Aphid AAMs beneath the fuselage and larger AA-7 Apex AAMs beneath the wing. The most important NATO AAMs are the radar-homing Sparrow (top right) and the IR-guided Sidewinder (right). A USAF technician (below) works on an F-4E's APQ-120 radar.

generally weapons load is restricted by the need to carry drop tanks on the fuselage centerline or underwing.

The greatest number of tactical fighters in Frontal Aviation are now versions of the MiG-23 and MiG-27 family, codenamed Flogger by NATO. The MiG-23 is a variable-geometry wing, single-seat fighter which can successfully undertake a variety of roles, including interception, air superiority, battlefield interdiction and close air support. The initial production versions, designated Flogger-B, are primarily intended for the air-to-air roles with secondary ground attack capability. Early examples were powered by the Tumansky R-27 afterburning turbofan producing 22,500lbs thrust, but this was superseded by the 25,350lbs thrust R-29 in the mid-1970s. Maximum speed at sea level is Mach 1.1 increasing to Mach 2.2 at 36,000ft. Service ceiling is 55,000ft and combat radius 575 miles. Span is 46ft 9in (17 degrees sweep) reducing to 27ft 2in when the wing is fully swept (72 degrees) and maximum takeoff weight is 41,000lbs.

The MiG-23's High Lark radar is a considerable improvement over the MiG-21's 20 mile range Jay Bird, having a search range of more than 50 miles and limited 'look-down' ability. Armament initially comprised the AA-2 Atoll infra-red guided AAM and the radar-guided AA-2-2 Advanced Atoll. However these missiles have given way to AA-7 Apex medium-range, radar-guided AAMs and the IR AA-8 Aphid for dogfighting. Two AA-7s plus four AA-8s can be carried. Gun armament is the 23mm GSh-23 cannon and bombs or rocket pods for the ground-attack mission. A laser range-finder is mounted forward of the nose wheel for air-to-surface weapons ranging. Improvements to the Flogger-B have resulted in the Flogger-G, which is also primarily an air-to-air fighter. The most noteworthy development is the provision of a pulse-doppler radar which can engage targets flying below the fighter.

The MiG-23 is in no real sense a successor to the MiG-21, but rather has filled a gap by providing a multi-role capability roughly comparable to that of the USAF's F-4 Phantom. A replacement for the MiG-21 in the air superiority and air defense roles is expected to become fully operational around 1985. Variously identified as Ram-L and MiG-29, the new fighter is a twin-engined, twin-fin single-seater, similar in appearance to a scaled-down F-15 or MiG-25. Performance is likely to be in the order of a Mach 2.3 maximum speed at altitude and Mach 1.2 at sea level. The Ram-L/MiG-29 will have a 'lookdown/shootdown' capability and generally its performance will be similar to or better than that of the McDonnell Douglas F-18, except for range which is significantly less. If reports on this new Soviet fighter are substantially correct then the technological gap between NATO and Warsaw Pact aircraft will be rapidly closing by the end of the 1980s.

For a time during the 50s and early 60s it was believed that an air superiority fighter could rely on missiles alone. The C and D models of the Phantom had no internal gun armament and when combat experience in Vietnam showed this to be necessary they had to carry a podded 20mm cannon. This was unsatisfactory because the weapon was less rigidly fixed than on an internal mounting and accuracy suffered. Ammunition capacity was also less than most internal gun arrangements and the gun occupied a stores station which could more usefully be taken up by bombs or fuel. The deficiency was finally corrected with the F-4E and its built-in 20mm M61 cannon, capable of firing at 6000 rounds per minute.

Nevertheless missiles remain the fighter's primary weapon. During the air war in Southeast Asia AAMs accounted for 88 of the 137 North Vietnamese fighters claimed as destroyed. The tactical fighters' guns destroyed 40 MiGs, while seven were maneuvered into the ground and two claimed by defensive fire from B-52s' tail guns. Two of the missiles used in Vietnam are still in NATO service, albeit in improved versions: the AIM-7 Sparrow (50 victories) and the AIM-9 Sidewinder (33 victories).

The Sparrow is a medium-range missile, which can be effective up to 60 miles from the launch aircraft in its AIM-7F version. It makes use of semi-active radar homing, which means that the parent aircraft's radar must lock onto the target for the missile to guide onto it. In an active homing missile, such as the Hughes AIM-54 Phoenix carried by the F-14 Tomcat fleet defense fighter, the illuminating radar is carried in the missile itself. The AIM-9 Sidewinder makes use of infra-red

guidance which homes onto a heat source which the seeker head has been locked onto before the missile is launched. Effective range is 6-10 miles and the latest AIM-9L Sidewinder can be launched from 'all-aspects', rather than from the target's rear as with earlier Sidewinders. Infra-red guidance can be decoyed away from a target aircraft's engine exhaust by flares or a jamming beacon and a natural heat source such as the sun can deflect the missile from its target. However the system has the advantages of robustness and relative simplicity and because it does not need radar guidance it can be carried by virtually any aircraft for self-defense and by advanced trainers in an auxiliary air defense role. A replacement for Sidewinder is to be produced by a European consortium and an active radar successor to Sparrow is under development in the USA.

The Soviet counterpart of the Sidewinder is the AA-2 Atoll, which makes use of infra-red guidance. An advanced version, the AA-2-2 probably exists in both IR and semi-active radar homing versions. Range is 3.5 miles, but Atoll's reliability has been poor when used in combat in Vietnam and the Middle East wars. The AA-7 Apex and AA-8 Aphid are supplanting the earlier missile, both types arming the MiG-23 and the AA-8 being carried on MiG-21s. There are two versions of both missiles, one semi-active radar homing and the other using IR guidance and both types are often carried by Soviet fighters. The medium range (20 miles) AA-7 is generally comparable to the Sparrow, while AA-8 is a small, maneuverable dogfighting weapon of some four miles range.

Radar is important in the air superiority mission, not only for illuminating targets for semi-active radar homing missiles, but also for target acquisition and ranging. The Hughes APG-65 radar fitted to the US Navy F-18 Hornet illustrates the capabilities of a modern tactical fighter's radar. It is a multi-mode radar, providing weapons aiming information for use with the 20mm cannon, Sparrow and Sidewinder AAMs, or guided and ballistic air-to-ground weapons. The radar modes applicable to the air-to-air mission include velocity search for picking up targets at a range of up to 80 nautical miles. As the target closes the radar can provide range information using the range-while-search mode. Thereafter the radar switches to single target track for guidance of Sparrow missiles, while 'fire-and-forget' missiles (active radar homing or infra-red guided) allow the radar to use the track-while-scan mode, providing position information on targets while still searching for others. An automatic priority is allocated to targets by the radar and up to ten can be tracked at once. It is also possible to assess the strength

An important US precision-guided munition is the TV-guided Maverick (inset top). The A-7Ds (left) with ventral dive-brakes extended carry Mk82 bombs. The primary weapon of the A-10A is its 30mm cannon. A-10As of the USAF's 355th TFW are shown (inset left).

of a raid by radar beam sharpening. For dogfighting targets are automatically acquired at ranges from ten miles down to 500ft. The radar can scan a narrow beam above and below the fighter, look forward along the aircraft's centerline, or scan the field of view of the HUD. A gun director mode indicates a target's relative motion during dogfighting.

Even in the age of the missile something akin to the traditional dogfight will take place. Combat maneuvers generally have the aim of placing the attacking fighter in a favorable firing position astern of his opponent or conversely of maneuvering out of an enemy's cone of fire and turning the tables on him. A number of modern developments have complicated the dogfight. Some missiles such as the AIM-9L Sidewinder can be fired from virtually any angle off the target and still successfully guide onto it. Similarly developments in flying control systems enable a fighter to pitch up into a favorable firing position in a maneuver that would render a conventional aircraft uncontrollable. Consequently simply keeping an enemy aircraft out of the 'six o' clock' position will no longer be any guarantee of safety.

Knocking out the enemy air force on his own airfields is an attractive idea, but the Israeli success in 1967 is unlikely to be repeated unless tactical nuclear weapons are used. Nonetheless, an enemy's air effort can be seriously blunted by a counter-air campaign against his airfields. This is especially true if, as in the case of NATO, he is somewhat short of airfields and those he has are crowded with warplanes, their support equipment and supplies.

Airfield attack will be one of the priority tasks for interdictor aircraft such as the F-111, Tornado and Fencer. Although they may carry conventional high explosive bombs or tactical nuclear weapons, it is more likely that their armament will be one of the specialized weapons developed for airfield attack. France's Matra Durandal is a direct descendant of the 'dibber' bombs used during the Six Days War. The weapon weighs 430lbs and up to 11 can be carried by the Jaguar, while the F-15E can lift a maximum of 22. The bomb is released over an enemy runway by a low-flying aircraft flying at up to 550 knots. Two braking parachutes are then deployed to decelerate the bomb and to pitch its nose downwards. The parachutes are then jettisoned and a rocket motor accelerates the weapon to a rate of some 850ft per second. It penetrates up to 16in through concrete and the 33lb warhead explodes after a one second delay. The French BAP 100 is a smaller weapon (80lbs weight) than Durandal, but works on the same principle.

The effects of a well-placed Durandal can be most satisfactory. A crater 16ft across and 7ft deep is made in the immediate area of the explosion, with slabs of concrete raised out to 50ft from the crater and further cracking of the surface over a greater distance. The area affected may be 300 square yards. Repair of this damage could take up to a day and it will be further hindered if the load of Durandals is supplemented by delayed-action anti-personnel cluster bombs.

Britain and West Germany have elected to develop large dispensers for carriage beneath the aircraft which eject numerous submunitions over the target. The RAF's JP233 was originally to have been a joint venture with the USAF, but the Americans decided not to continue with the project because they were worried about the vulnerability of the carrier aircraft to airfield defenses. JP233 carries both runway-cratering and anti-personnel bomblets in the same dispenser and two dispensers can be carried by a Tornado beneath the fuselage. West Germany's MW-1 is a similar concept for arming Luftwaffe Tornados and F-4 Phantoms. Up to 224 bomblets can be carried by a single large dispenser and these can be either anti-armor, anti-personnel, or runway-cratering submunitions. This last type has a double charge warhead to crater the runway and then explode beneath its surface. More than 1500ft of runway can be damaged by the MW-1 during a single pass.

After withdrawing from the JP233 program, the USAF has evaluated a number of anti-airfield weapons, including Durandal, BAP 100 and the Canadian CRV-7 high velocity rocket with a 16lb warhead. However none of these has a true stand-off ability and development work in the USA is concentrating on submunition-dispensing ASMs for airfield attack. The AGM-109H variant of the Tomahawk cruise missile, which can be carried by B-52s and F-16s, will deploy some 60 submunitions. Similar applications of the Phoenix and AGM-86 ALCM and the British P4T (a cruise missile development of the Sea Eagle) are being considered. Soviet anti-airfield weapons are thought to include dibber bombs, cluster weapons and a bomb intended for hardened aircraft shelters.

During the 1970s much was done to reduce the vulnerability of military airfields. Important airfields are now defended by concentrations of AA guns and SAMs, with possibly a combat air patrol of fighters overhead. A general 'toning down' of aircraft, equipment, buildings and runways themselves make airfields and targets on them difficult for an attacker to pick out from the surrounding terrain. Aircraft, support equipment, munition and fuel stores are not only dispersed, but are also often protected by shelters able to withstand anything but a direct hit from a heavy bomb.

The most vulnerable target on a military airfield is still the runway. Tactical aircraft require up to 4400ft of runway on takeoff, although this distance can be reduced by carrying a smaller load. The high-performance F-15, operating as an interceptor, can leave the ground within 900ft, while even the Harrier operating in the STOVL (short take-off vertical landing) mode to maximize its warload will need some 1200ft. The transport aircraft so vital for NATO reinforcement and resupply are much more demanding. The C-130

Hercules needs 4700ft, while the C-5A Galaxy's takeoff run is 8000ft. Landing runs are generally shorter, ranging from the Harrier's nil to the F-4's 3800ft.

Various measures can be employed to guard against an airfield's operating surfaces – runways, taxiways and hardstanding – being rendered completely unusable. One approach is to provide rapid repair teams and to stockpile the materials that they will need. USAF Red Horse units make use of aluminum matting patches and filler materials to repair damage and, providing unexploded ordnance can be rapidly cleared, they can make repairs within a few hours. It is unlikely that a single attack will knock out all of an airfield's operating surfaces. Aircraft such as the Jaguar are able to fly from taxiways or even grass strips when the main runway is unavailable. Even warplanes with less favorable rough-field characteristics can make use of undamaged strips of runway – perhaps finishing the takeoff run on grass when the wing has begun to take some of the aircraft's weight. However, such operations will usually have to be at a reduced takeoff weight, hence effectiveness will be blunted. Landing back on a damaged airfield will be even more difficult.

A US Marine Corps A-4 Skyhawk uses jet assisted take-off when operating from a SATS (short airfield for tactical support).

Dispersed operations away from the main airfield offer one solution to the problem. Such tactics are especially associated with the V/STOL Harrier, which in wartime will operate from camouflaged sites well away from regular airfields. However, any warplane with a degree of STOL (short take-off and landing) capability can operate from such sites as suitable lengths of major highway. The disadvantage is the difficulty of providing remote sites with adequate supplies of fuel and munitions. Any major servicing will also be extremely difficult without well-equipped workshops. Finally if a dispersal site is discovered and attacked by enemy aircraft or heliborne commandos, its defenses could be far less effective than those of a permanent airfield.

The Swedish air force, which has for many years trained in flying from suitable stretches of roadway, aims to provide its planes with a large number of 1640ft stretches of road, which can be reached by taxiway from its regular air bases. This makes it virtually impossible for an anti airfield campaign to succeed and yet dispersed aircraft remain within reach of their base. Other methods include deploying to civil airfields, or (as favored by the Soviet Union) making use of a large number of rough airstrips in addition to the well-equipped regular bases. The US Marine Corps has a specialized requirement for establishing an airstrip ashore as soon as possible during an amphibious assault. Consequently it has developed a unique method of operating, using a SATS or short airfield for tactical support. This is in effect a shore-based aircraft carrier, using catapult or JATO (jet-assisted takeoff) and arresting gear for landings. The Marines also fly the Harrier and have experimented with a land-based ski-ramp (similar in concept to that used by the Royal Navy's aircraft carriers) to achieve a short takeoff with a fully-laden aircraft.

Air defenses consist not only of interceptor aircraft, but also of ground based systems such as SAMs, AA artillery and their associated radars and command centers. It is therefore an important part of the air superiority mission to neutralize these. The USAF has gained a great deal of experience in defense suppression tactics, which it codenames Wild Weasel. The need for active defense suppression varies somewhat according to the penetration tactics adopted by the attacking force. The USAF, like most other NATO air forces, flies tactical strike/attack missions at medium level, covered by fighter and ECM escorts, and therefore required active suppression of ground based defenses. An alternative approach, adopted by the RAF, is to operate at high speed and low level, thus giving ground defenses only a fleeting opportunity to come into action.

Soviet ground-based air defense systems include AA guns ranging in size from 14.5mm up to 130mm, with the ZSU-23-4 quad-barrel 23mm gun and the twin 57mm ZSU-57-2 representing a particularly serious threat to low flying aircraft. SAMs vary greatly in operating altitudes and mobility, but in general provide good air defense cover at all altitudes and in all weathers. SA-4 medium to high altitude missiles are deployed at army and front levels, while divisions have SA-6, SA-8 and SA-11 SAMs operating at low-to-medium altitudes and the regiments have low-altitude SA-9 and SA-13 missiles. In addition to the ground forces' missiles, deeper penetrating tactical aircraft will have to cope with the Air Defense Force's SAM force which currently numbers some 10,000 launchers.

The tactics evolved for dealing with SAMs are an outgrowth of the air war in Southeast Asia. In 1965 'Iron Hand' bombing strikes were initiated against SAM launching sites and later that year specialized 'Wild Weasel' F-100F aircraft were introduced to deal with the SAMs while the strike force went for the main target. The F-105F was also assigned to this role and from this aircraft evolved the F-105G. This two-seat variant of the Thunderchief was fitted with special electronic equipment which enabled it to detect and home in on the emissions of SAM guidance radars. The SAM sites could then be attacked by anti-radiation missiles, which home onto the radar signals, or by conventional bombs, rockets and CBUs.

Veteran F-105Gs still fly with the Air National Guard, but the front-line Wild Weasel units now operate the F-4G version of the Phantom. The F-4G is a modified F-4E fighter, 116 of which are to be converted for the defense suppression role. It is fitted with the APR-38 radar homing and warning system, which detects AA gun and SAM radar emissions. The system's computer will provide the position of the hostile transmitter and identify it as an AA radar, or SA-2, 3, 4, etc. as appropriate. The system will also assign threat priorities for up to 15 different radars. The computer will then provide weapons release data, or execute an automatic bombing run if desired.

The weapons carried by the F-4G include the AGM-45 Shrike and AGM-78 Standard anti-radiation missiles and the new AGM-88 HARM, which is under development for the defense suppression mission. The electro-optically guided AGM-65 Maverick provides a useful answer to such simple countermeasures as shutting down the transmitter. Other air-to-ground ordnance may include free-fall bombs, PGMs and CBUs. The F-4G can defend itself against enemy aircraft as it can carry three Sparrow missiles (the position of the fourth Sparrow being taken up by an ECM jamming pod) and AIM-9 Sidewinders. However one retrograde step is the deletion of the internally-mounted gun, as its position is needed for radar-warning antennae.

There can be little doubt that the Wild Weasel mission is one of the most hazardous in the tactical fighter's repertoire. In Vietnam it was often possible to outmaneuver an SA-2 if its launch was detected in time and one veteran of that conflict recorded that he would rather deal with SAMs than fly through small arms fire, as 'if you can see a missile, you can always beat it!' Whether this remains true of later systems than the 1950s-technology SA-2 is open to doubt. It may be that the defense suppression mission will become too dangerous for the manned aircraft and be taken over by remotely piloted vehicles (RPVs).

The most effective stand-off weapon of the Wild Weasels is the anti-radiation missile. The AGM-78 Standard ARM has a speed of Mach 2 plus and a range of 15 miles, while the AGM-54 Shrike's range is some ten miles. Both weapons were used operationally in Southeast Asia and are due to be replaced by the AGM-88 HARM (high-speed anti-radiation missile), which has an increased range, higher speed and a greater frequency coverage.

There is more than one way of skinning a cat and, if a direct assault on enemy air defenses can be costly and hazardous, then the subtler techniques of deception can often do the same job just as effectively. Electronic countermeasures can affect virtually all aspects of air operations, but electronic warfare (EW) is especially significant in the tactical air war and a number of specialized EW aircraft have been developed to provide ECM escort for strike/attack forces.

In general terms ECM can be either active or passive. Passive measures include the Stealth techniques already discussed and the use of radar warning receivers to alert an aircraft's crew to a hostile threat. The active approach seeks to jam enemy radars and communications, or to confuse the defenders by deception. The widespread use of ECM from World War II onwards has led to ECCM (electronic counter countermeasures), which seeks to make air defense radars immune from enemy jamming by such techniques as frequency agility. This involves a rapid and random switching of the frequency on which radars or radio communications sets operate to make the task of jamming formidably difficult.

The USAF is currently developing an ECM escort aircraft based on the F-111A airframe. Known as the EF-111A, this airborne tactical jamming system makes

A Soviet Yak-28 Brewer-E electronic warfare aircraft pictured in service with Frontal Aviation.

use of the ALQ-99 electronic warfare equipment, which was originally produced for the US Navy's EA-6B Prowler. However the EF-111A's ECM operation is more automated as the EF-111A carries only one systems' operator in contrast to the Prowler's three. Essentially the operator will monitor an automatic response to electronic threats, although manual override is also possible. The EF-111A carries jamming transmitters in an underfuselage canoe fairing, with the receiver antennae mounted in a large pod atop the tailfin. Essentially similar to its tactical fighter-bomber counterparts, the aircraft has sufficient performance to accompany a strike force, but it could equally operate as a 'stand-off' jamming system.

The US Navy's EA-6B Prowler is a much earlier system, which first entered squadron service with the US Navy in 1971. In essence it is a modification of the A-6 Intruder attack bomber, with a four-man crew in an enlarged cockpit. Its jamming system is the same as that fitted to the EF-111A, but instead of being internally mounted the transmitters are carried in self-contained underwing pods. Up to five pods can be carried, each of which has two jamming transmitters, and the Prowler also carries a communications jamming set. Although it is a carrier based naval aircraft, the Prowler's mission otherwise differs little from that of land-based ECM aircraft.

There has been very little published in the West about the Soviet Union's ECM equipment and capabilities. However this is one of the areas of military technology in which the Soviet Union is claimed to excel. It is thought that most Soviet warplanes carry internally-mounted ECM systems and a number of specialized EW aircraft have been identified. Frontal Aviation's standard ECM escort is the version of the Yak-28 known to NATO as the Brewer-E. The Yak-28 is a twin-engined, two-seat warplane which undertakes

a variety of missions including attack, interception and reconnaissance. The Brewer-E is based on the attack version, with the weapons' bay housing ECM equipment. Its performance – Mach 0.85 at sea level, increasing to over Mach 1 at medium altitudes – suggests that it will accompany strike formations as ECM escort. The Soviet stand-off jamming system is carried by the Cub-B variant of the Antonov An-12 tactical transport. This is similar to the USAF's EC-130 ECM variant of the Hercules. The advantage of modifying a tactical transport is that there is plenty of room to accommodate operators and equipment. This must be offset against the vulnerability of such an aircraft and so these systems must operate under the cover of friendly air defenses.

Specialized ECM aircraft are only available in relatively limited numbers when compared with the hundreds of tactical aircraft which will operate over the European battlefield. Therefore most warplanes are fitted with their own ECM equipment, which varies in complexity and effectiveness according to the level of opposition anticipated. The first requirement is to detect a threat from hostile radars and this is provided by radar warning receivers. Such equipment may simply indicate the direction and nature of a single transmission. Alternatively the equipment may be able to detect, identify and show the position of a dozen or more threats simultaneously. It is also possible for the system to allocate a priority to each threat and automatically to initiate appropriate countermeasures.

Jamming and other ECM equipment is usually carried by a fighter in a self-contained pod mounted on one of the weapons' hardpoints. This has the disadvantage of occupying a stores station which otherwise could be used for ordnance, but it also allows more flexibility than an internal ECM system. The pod can be removed if ECM support is not required and replacing an obsolescent pod is simpler than refitting an internal ECM suite. Most Western air forces use pod-mounted jammers, but advanced fighters like the F-14 and F-15 have internal systems. The Belgian air force alone among NATO operators of the F-16 has elected to fit the Loral Rapport III internal ECM system to this fighter, while other operators favor ECM pods.

The crudest method of jamming a hostile radar is by 'noise', in other words by blotting out the radar's signal by a stronger transmission on the same frequency. This can be spot jamming on a single frequency or barrage jamming over a wide range of frequencies. Noise jamming is relatively simple to implement, but it is also fairly easily countered by changing frequencies or by frequency agility. A more subtle approach seeks not to jam the enemy's signals, but to distort them. This is achieved by receiving the signal and retransmitting it in a modified or delayed form. Alternatively the enemy can be confused by generating a cluster of false target returns in which the genuine return is lost. Finally most tactical aircraft will be fitted with chaff and flare dispensers. A cloud of chaff – minute strips of aluminum foil often cut in lengths to match the radar wavelength – will break a radar lock on its target, while IR flares can decoy an IR missile from the aircraft's engine exhaust. There are also more sophisticated electro-optical jammers intended to decoy heat-seeking missiles from their target.

Air power's most direct contribution to the land battle is the close air support (CAS) mission, which provides firepower for ground forces in contact with the

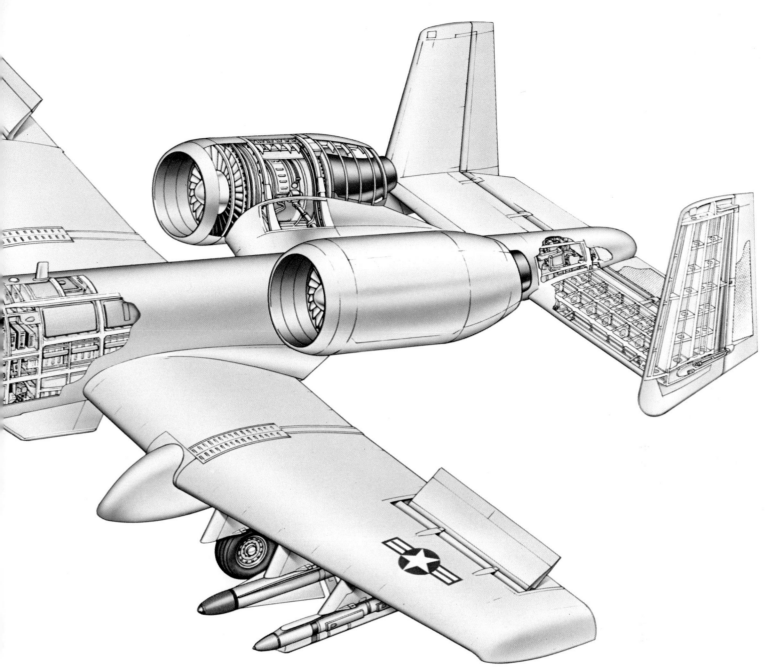

The A-10A Thunderbolt II, otherwise known as the Warthog, is a specialized anti-tank and close air support aircraft (cutaway artwork). The RAF Buccaneer S Mk 2Bs (left) belong to No 208 Squadron based at Honington in the UK.

enemy. Targets include armored and soft-skinned vehicles, troops and field fortifications. Because of the dangers of attacking one's own forces in error, positive control of CAS aircraft is of the greatest importance. This can sometimes best be achieved by an airborne forward air controller, orbiting the battle area in a light aircraft, who can mark ground targets with smoke rockets and direct the bombing runs of the attack fighters. There will be many attractive targets to the rear of the enemy's forward troops and so battlefield interdiction will also be undertaken by tactical air forces.

Close air support operations are fraught with many difficulties. The problem of identifying hostile forces and avoiding attacking friendly troops will be formidable in the smoke and confusion of battle – hence the need for forward air controllers on the ground and in the air. There follows from this the need for effective communications between air and ground. Laser designation of targets will help the delivery of precision-guided munitions, but their advantages in CAS work are not as great as may be supposed. Many air forces favor area weapons such as cluster bombs which do not require highly accurate deliveries. Because of the heavy concentration of hostile AA and SAMs the battlefield will be a dangerous place for the attack aircraft and weapons which can 'kill' in the first pass at the target are highly desirable. Drones or remotely piloted vehicles are theoretically attractive for these hazardous sorties, but they lack the inherent flexibility of the manned aircraft. Finally CAS aircraft will be operating in the same area as army attack and scout helicopters and close co-operation between the two forces is both prudent and highly desirable.

The most formidable target on the battlefield is the tank and the USAF's standard attack aircraft, the Fairchild A-10A Thunderbolt II (unofficially dubbed the Warthog) has been specially developed to deal with enemy armor in the Central European theater. It is a relatively simple, single-seat aircraft intended to be available in large numbers (the A-10A-equipped 81st TFW is the largest fighter wing in the USAF with six 18-aircraft squadrons) and survivability is one of the keynotes of the design. The A-10A is relatively large, spanning 57ft 6in with a length of 53ft 4in and a maximum takeoff weight of 47,400lbs. Power is provided by two General Electric TF-34-GE-100 turbofans of 9065lbs thrust each, giving a maximum combat speed with warload of 443mph. Operational radius, allowing a loiter time over the target area of two hours, is 288 miles. Maximum warload is 16,000lbs.

The built-in armament of the A-10A comprises a 30mm GAU-8/A seven-barrelled rotary cannon, with a rate of fire of either 2100 or 4200 rounds per minute according to the pilot's selection. Ammunition load is 1350 rounds and so the gun is fired in short (typically two second) bursts. Each ammunition round has an armor-piercing cone of depleted uranium, backed by an incendiary charge, and it is reckoned to knock-out a tank with hits on the side or rear at a distance of 6000ft. Beyond gun range a favored weapon is the TV-guided Maverick ASM, up to six of which can be carried. Alternative armament loads may include bombs, CBUs, or laser-guided bombs.

The A-10 has been designed to absorb considerable battle damage and be able to return to base. The cockpit area is protected by a titanium shield which can withstand hits from a 37mm shell. Fuel tanks and their associated 'plumbing' are protected and are self-sealing and the engines are mounted high on the rear fuselage so that they will be partly shielded from groundfire. The ammunition tank is armored and hydraulic flying control systems are duplicated with manual reversion. Finally the aircraft's structure is particularly strong and it is claimed that the A-10 can lose one engine, half a tail and two-thirds of one wing and still remain in the air.

When operating over the European battlefield the A-10 relies on its low operating altitude and its good maneuverability for safety. Flying below 100ft it will be very difficult for a hostile fighter to pick up and if intercepted the agile A-10 will be able to turn into the fighter's attack forcing him to overshoot. Secondly, the A-10 will make use of ground masking to avoid groundfire, and can use evasive 'jinking' maneuvers during its firing passes. The aircraft will operate in close conjunction with army helicopters and artillery and can

An A-37B Dragonfly light attack aircraft (above) of the USAF. The AC-130 (top) carries a broadside armament of 20mm cannon and 40mm Bofors.

rely on these to provide suppressive fire against enemy ground defenses.

The most serious criticism of the A-10A is its lack of an all-weather capability. Because of the well-known vagaries of the European climate, there is a danger that the A-10 force may be grounded by bad weather and in winter daylight hours are in any case limited. A two-seat night and adverse weather variant of the A-10 was produced to cope with these conditions, but after evaluation it was decided to improve the single-seater. This will be achieved by fitting the LANTIRN system, which consists of a pod-mounted FLIR for navigation and targeting and a radar pod for terrain following. It remains to be seen whether a pilot alone can cope with the workload involved in all-weather attack sorties, especially as the demands of daytime low-level operation are already very high. However the economies of the LANTIRN system over the two-seat A-10 are self-evident.

The Warthog's predecessor in front-line service with the USAF, the Vought A-7D Corsair II, was an altogether more conventional aircraft. Although it has now retired from the regular squadrons, the A-7 remains an important aircraft with the Air National Guard, equipping 14 squadrons, and it is still the US Navy's standard shipboard light attack aircraft. The A-7D spans 38ft 9in and length is 46ft 2in, with a maximum takeoff weight of 42,000lbs. It is powered by a 14,250lbs thrust Allison TF-41A-1 turbofan, which gives a maximum speed of 663mph at 7000ft and a combat radius of 556 miles. The built-in armament comprises a 20mm M61A-1 rotary cannon with 1000 rounds of ammunition and the external ordnance load is up to 9500lbs. In addition two Sidewinder AAMs can be carried on fuselage pylons for self-defense. The A-7D's capabilities as an attack aircraft are due to a highly accurate navigation and weapons delivery system, which provides the pilot with navigational information and weapons release instructions projected onto a cockpit HUD. A central computer is fed with information from an inertial navigation set, doppler radar and attack radar.

Close air support is also undertaken by the USAF's multi-role F-4 and F-16 fighters. An altogether more specialized aircraft is the AC-130 gunship, developed for night fire-support in Vietnam, which is still in service with the regular air force and AFRES. A conversion of the standard C-130 tactical transport, AC-130s are fitted with a broadside armament of 7.62mm and 20mm multi-barrelled guns, supplemented in the later AC-130H model by a 40mm cannon and a 105mm howitzer. The aircraft is also fitted with an array of target acquisition sensors, including radar, low-light TV, IR detection equipment and night observation starlight scopes. A computer instructs the pilot to orbit his target at the appropriate angle of bank to concentrate the cone of fire from his side-mounted armament onto the target. Although a very effective system in Southeast Asia, the AC-130 is vulnerable to interception and can only operate in areas where the enemy air threat has been eliminated. An even more unlikely participant in World War III is the Piper Enforcer, basically a turboprop-powered derivative of the World War II P-51 Mustang fighter, which the USAF is to evaluate in the close air support role.

Forward air control of CAS sorties by an experienced airman accompanying the ground forces became a standard operational procedure in Western air forces during World War II. This not only ensures that enemy positions are clearly marked for the CAS aircraft, but also that the army commander's requirement has been realistically assessed by an airman with recent combat experience who understands the limitations and problems of CAS. As a logical extension of this practice, forward air control from an aircraft over the battlefield is now widespread in the US forces.

The aircraft used for FAC include the Cessna 0-2 development of the twin-engined Skymaster light aircraft, with underwing hardpoints for target marking rockets. This aircraft has a maximum speed of 200mph and a range of more than 1000 miles. Although good visibility for ground observation is more important in a FAC aircraft than performance, a more capable aircraft than the 0-2 is really required for the European battlefield. This is available in the guise of the Rockwell OV-10A Bronco, a two-seat twin-turboprop counterinsurgency aircraft developed for service in Vietnam. The Bronco combines excellent crew visibility with a good performance (maximum speed 280mph) and the ability to carry a weapons load of 3600lbs in addition to four built-in 0.30in machine guns, so that it can itself undertake light attack missions. A further increase in the performance of FAC aircraft has resulted from converting jet aircraft to this task. The USAF has modified Cessna A-37 attack aircraft to OA-37 configuration for service with reserve units and the US Marine Corps has similarly converted two-seat Skyhawks as OA-4Ms.

Undoubtedly the most interesting and original CAS aircraft serving with the United States' NATO allies (and also with the US Marine Corps) is the RAF's B Ae Harrier V/STOL fighter, the only land-based fighter of its type in operational service. The Harrier's vertical takeoff capability enables it to operate from dispersed sites close behind the battlefield, so that requests for air support can be quickly met. It is also independent of vulnerable and conspicuous permanent airfields, although as already noted the demands of resupply and maintenance of a relatively complex aircraft present other problems. When operating with a full warload the Harrier must use a short-takeoff run, although with ordnance expended it can land vertically. In practice this restriction has not hampered its operation from dispersed camouflaged sites and this capability represents a practical yet unconventional response to the problems of airfield vulnerability.

Three front-line RAF units currently operate the Harrier, Nos 3 and 4 squadrons at Gutersloh, Germany and No 1 squadron at Wittering in the UK. The latter squadron is assigned to NATO's ACE Mobile Force with a reinforcement role for NATO's flank areas. The operational US Marine Corps units are VMA-231 and VMA-542 at Cherry Point, North Carolina, and VMA-513 at Yuma, Arizona. The Marines pioneered the use of the Harrier (designated AV-8A in US service) in air-to-air combat, developing the use of vectored-thrust in forward flight, or 'viffing', to force conventionally powered fighters into overshooting.

The RAF's Harrier GR Mk 3 is powered by a 21,500lbs thrust Rolls-Royce Pegasus 103 vectored thrust turbofan, which gives a maximum speed of Mach 0.95. Maximum weight for vertical takeoff is 18,000lbs, increasing to 23,000lbs for a short takeoff. The Harrier is small, spanning 25ft 3in and is 45ft 8in long. Tactical radius is 400 miles and warload is 5000lbs of ordnance, plus two 30mm Aden cannon. Navigational equipment includes an inertial nav/attack system and a laser ranger and marked target seeker is fitted.

The AV-8B (Harrier GR Mk 5) will offer a significant improvement in payload/range over the present machine. It is a developed version of the original Harrier, built by McDonnell Douglas in co-operation with British Aerospace for the US Marines (336 required) and the RAF (60 required). It has a larger wing with leading-edge root extensions to improve the rate-of-turn. Other refinements include a raised cockpit

with better visibility, six underwing stores stations (the AV-8A has four), the Hughes Angle Rate Bombing System with TV and laser spot tracking and a pod-mounted 25mm rotary cannon. The first AV-8Bs should become operational in 1985.

Complementing the Harrier in RAF service, the SEPECAT Jaguar (built jointly by Britain and France) operates primarily in the battlefield interdiction role, but it can also undertake close air support, tactical reconnaissance, airfield attack and tactical nuclear strike. Jaguar was designed from the outset for low-level operation, with the capability of destroying its target during the first pass. Powered by two Rolls-Royce Turboméca RT172 Adour turbofans of 7140lbs thrust with reheat, Jaguar has a maximum speed of Mach 1.1 at 1000ft. Range with external fuel tanks flying a lo-lo-lo sortie is 564 miles. It spans 28ft 6in and is 50ft 11in long, with a maximum takeoff weight of 32,600lbs. It can carry an external weapons load of up to 10,000lbs on five hardpoints and built-in armament is two 30mm Aden cannon. An overwing mounting for two Sidewinder or Matra Magic AAMs has been tested, but they are not fitted to RAF or French air force Jaguars.

The heart of Jaguar's avionics is the inertial NAVWASS (navigation and weapons' aiming subsystem), which provides navigational and weapons aiming data on a HUD. It also drives a moving map display in the cockpit. Five squadrons in RAF Germany are equipped with the Jaguar and three fly the type from the UK. Two Jaguar squadrons operate in the tactical reconnaissance/strike role. In the Armée de l'Air the Jaguar equips nine escadrons of the Force Aérienne Tactique.

Close air support aircraft do not need to be as complex as the Jaguar. Advanced trainer aircraft, such as the Macchi MB326, Fouga Magister, Saab 105, B Ae Hawk and Franco-German Alpha Jet, have a limited capability in combat roles. This can be enhanced by modifying the trainer into a light attack aircraft, as in the case of the A-37 variant of the USAF's Cessna T-37, or the Strikemaster version of the Jet Provost. Any advanced training aircraft used for weapons instruction can assume a secondary combat role if necessary. West Germany has put its 175 Alpha Jet trainer/light attack aircraft into first-line service with three of its Jagdbombergeschwadern, whereas France uses the type in the advanced training role.

For many years the Soviet air force neglected close air support, preferring to rely on artillery and surface-to-surface missiles for fire support, while Frontal Aviation provided fighter cover. However during the 1970s as part of the general increase in the effectiveness of Soviet tactical air power the ground support capabilities of Frontal Aviation were significantly upgraded. However control of attack aircraft is not as flexible as in NATO, with army requests for air support being filtered through air force liaison officers and requiring approval at many levels of command. Although this procedure can be shortcircuited in an emergency, its speed cannot compare with that of the FAC system employed by Western forces. The reluctance to dele-

gate authority to junior levels of command is a recurring weakness in the Soviet system.

The most capable attack fighters in Frontal Aviation service at present are the MiG-27 Flogger-D and -J versions of the MiG-23 air-to-air fighter. A major change from the MiG-23 is the recontouring of the nose to improve the downward view from the cockpit. The air interception radar is replaced by a small ranging radar, a laser rangefinder and doppler radar to measure ground speed. The MiG-23's variable engine inlets are replaced by large fixed inlets and the afterburning is simplified. These modifications are designed to increase range at medium and low altitudes at the expense of Mach 2 performance, unnecessary for the attack mission. The undercarriage is modified for rough field operation, and a rocket-assisted takeoff pack can be fitted. Built-in armament is a single 23mm rotary cannon and the weapons load is 6600lbs. Performance includes a maximum speed of Mach 1.1 at sea level and a combat radius of 575 miles. A version of the MiG-23 known as Flogger-F is fitted with the new nose but lacks the other features of the MiG-27. It has been supplied to Warsaw Pact satellite air forces and exported to Africa and the Middle East.

The Sukhoi Fitter family of strike fighters offers another instructive example of the Soviet practice of progressively modifying an existing design to meet new requirements. The Su-7 Fitter-A is a swept-wing fighter-bomber, which served in quantity with Frontal Aviation in the 1960s and has not yet retired. It is a stable aircraft at low level, where it attains a maximum speed of Mach 1.1. However the Su-7 lacks effective navigation and weapons aiming avionics and its payload/range characteristics are poor.

Rather than replace the Su-7, the Soviet Union has modified the Fitter's wing with variable-geometry outer panels. These give some of the advantages of a true variable-sweep wing, but avoid the associated aerodynamic problems. The new aircraft, designated Su-17 (Fitter-C), was also fitted with a more powerful engine, the 24,250lbs afterburning AL-21F. It is undoubtedly an improvement over its predecessor, offering a better airfield performance and tactical radius, with nearly double the Su-7's warload. The later Fitter-D has a terrain avoidance radar and laser ranger and marked target seeker installed, thus improving another fault of the Su-7. Fitter-G is a further advance in this respect and probably carries laser-guided PGMs.

The Su-17 spans 45ft with wings spread full-forward, reducing to 34ft 6in when fully swept and overall length is 58ft. Maximum takeoff weight is about 37,500lbs. Maximum speed at low level is Mach 1.1 and tactical radius 300 miles. Two 30mm cannon are mounted in the wing roots and ordnance load, on six or eight external hardpoints, is up to 7500lbs. AA-2 Atoll AAMs can be carried for self-defense and auxiliary fuel tanks are usually carried.

A replacement for the Su-17, the Ram-J provisionally identified as the Su-25, is a specialized close air support aircraft. It is a single-seat, subsonic aircraft powered by two turbojets mounted in the wing roots. A multi-barrel cannon is fitted and there are 10 underwing ordnance pylons. According to American reports, the Ram-J is currently undergoing operational evaluation with the Soviet forces in Afghanistan. The inevitable comparisons with the USAF's A-10 are somewhat misleading as the Soviet aircraft is both smaller and lighter.

4. WAR AT SEA

A Super Etendard strike fighter of the French navy's Aéronavale prepares to catapult from its carrier.

War at Sea

Control of the seas – and equally importantly of the air space above them – will confer many benefits. The oceans are vital supply routes for the movement of military forces and their equipment, much of which is too bulky to be carried by air. The sea lanes give access to the raw materials and foodstuffs on which the industrialized states of Western Europe (and to some extent the USA also) depend for their very existence. Finally warships can intervene in the land battle by launching carrier air strikes or by mounting an amphibious assault. Naval aviation is an integral element of sea power and the aircraft carrier has become the capital ship of the modern age, with capabilities unequalled by any other warship.

The great advantage of the modern carrier is its mobility, combined with the flexibility of its air wing. A nuclear powered carrier can in theory remain at sea almost indefinitely, as its reactor fuel only need be replaced after long intervals (varying from 4-20 years). However the need to replenish aircraft fuel stocks, munitions and supplies of food are obvious limiting factors, as is the need to provide the crew with rest and relaxation. Furthermore the carrier's escorts may not be nuclear-powered and so the carrier group's endurance will be consequently reduced. The aircraft carrier is expensive in terms of the cost of the ship itself and of her specialized aircraft and in its demands for highly-trained manpower. Nevertheless it may provide the only base for tactical air power in an area where there are no friendly airfields ashore. The range of operations that can be undertaken by a typical carrier air wing is wide, embracing air defense, fighter escort, all-weather attack, airborne early warning, anti-submarine warfare, reconnaissance, electronic warfare and air refueling.

At present the US Navy is the only large-scale operator of aircraft carriers, although the Soviet Union is beginning to build this class of ship. There were 12 aircraft carriers available for deployment with the US fleets early in 1982. A 13th, the nuclear-powered *Carl Vinson* was commissioned that year and the USS *Coral Sea*, previously earmarked for relegation to training duties, was retained as an active carrier. A further nuclear-powered carrier, the *Theodore Roosevelt*, is scheduled to join the fleet in 1986 and two further nuclear ships are required for delivery in 1990/91. It is hoped to build up a force of 15 carriers, with the eight conventionally-powered ships of the *Forrestal* and *Kitty Hawk* Classes undergoing a life-extension program to add some 15 years to their anticipated service. The older *Midway* Class ships are likely to decommission in 1986.

A typical carrier air wing will embark nine squadrons, plus a photographic reconnaissance detachment, numbering in all 85-95 aircraft. The overall complement and the way it is made up may vary according to the class of carrier and the area of operations. However

This F-14A Tomcat of the US Navy's Fighter Squadron 32 The Swordsmen carries a full load of AIM-54A Phoenix air-to-air missiles (above). A number of Tomcats are to be fitted with the TARPS pod for reconnaissance, including this F-14A from USS Nimitz's VF-84 (right). An instructor mans the rear seat of an F/A-18 Hornet trainer assigned to VFA-125, Lemoore, Ca (top).

a nominal air wing comprises two fighter squadrons of 12 aircraft each, which could be F-14A Tomcats or older F-4J N, or S Phantoms. Two attack squadrons (24 aircraft) fly the Vought A-7E and a third has 10 Grumman A-6E Intruder all-weather attack aircraft. The Intruder squadron will also have four KA-6D tanker aircraft for air refueling missions. Anti-submarine warfare (ASW) is the responsibility of a ten-aircraft squadron of Lockheed S-3A Vikings and a helicopter squadron of six Sikorsky SH-3 Sea Kings. Airborne early warning is provided by four Grumman E-2C Hawkeyes and EW duties are undertaken by four EA-6B Prowlers. Photographic reconnaissance may be undertaken by a three aircraft detachment of F-14 TARPS (tactical airborne reconnaissance pod system) or Marine RF-4B Phantoms.

Because the aircraft carrier is such a valuable and scarce asset, defending it from air attack is of the utmost importance. Fleet air defense is therefore the primary mission of the air wing's two fighter squadrons, operating in conjunction with AEW aircraft and the close-in SAM defenses of the carrier and her escorts. As a bomber armed with a stand-off missile may launch its weapon 150 miles out from the task force and because the Soviet Union will probably employ saturation tactics – launching a co-ordinated series of attacks from different quarters in an attempt to swamp the defenses – a long range air defense system with the ability to engage several targets in quick succession is the ideal fleet defense fighter.

The Grumman F-14 Tomcat goes a long way towards meeting this exacting requirement, but it does have its shortcomings. For example the targets which it can engage simultaneously (up to six) must all be within the field of view of the F-14's AWG-9 radar until the Phoenix missile's own radar comes into action during the terminal phase of the missile's flight. A more general problem is that of identification, for whereas a radar can positively identify a friendly aircraft through the appropriate IFF (Identification, Friend or Foe) response, the absence of that response does not prove that the contact is hostile. It may be that a friendly aircraft's IFF equipment is faulty or damaged, alternatively a neutral aircraft could have strayed into the battle area. So the F-14 cannot always use its long range weapons (up to six AIM-54 AAMs). The aircraft accordingly can carry shorter-range Sparrow and Sidewinder AAMs and has a built-in gun armament of one 20mm M61A Vulcan rotary cannon with 675 rounds. The short-range armament would in any case be necessary for the secondary fighter missions such as attack force escort.

The Tomcat is a two seat, variable-geometry-wing fighter, powered by two Pratt & Whitney TF30 turbo-fans of 20,900lbs thrust with afterburning. The span with wings extended is 64ft 2in, length is 62ft and maximum loaded weight is 74,348lbs. At sea level maximum speed is Mach 1.2, which increases to Mach

F-4 Phantom crews of the US Navy receive air combat training with VF-171's Det Key West, whose F-4N and A-4E are shown (top). Many F-4 units including VF-84 (above) now fly the F-14 Tomcat. The remaining F-4 units and A-7 units will receive the F/A-18 (above right). A-7Es (right) of VA-146 carry FLIR pods.

2.4 at 49,000ft. The F-14A can climb to 60,000ft in 2.1 minutes and range for the combat air patrol mission is typically 765 miles.

The Hughes AWG-9 weapons control system comprises the radar, its computer and associated cockpit displays. Its detection range is two and a half times that of the F-4J's AWG-10 system. As a back-up, an infra-red target detection system is carried and a television system for visual identification at long range is being developed. Operating in the pulse-doppler mode the AWG-9 can detect a bomber at up to 170 nautical miles and a cruise missile at 62 miles.

The AIM-54A Phoenix can be launched at 63 miles range and because it has an active radar fire-and-forget guidance system, AIM-54As can be salvoed in quick succession. Under test conditions an F-14A has engaged six drone targets at once and achieved hits on four, one miss being attributed to a failure of the drone and the other to a fault in an AIM-54A.

The US Navy intends to equip 24 squadrons with the F-14 and by the beginning of 1982 about half of this force was in service. The Tomcat first deployed operationally aboard the carrier *Enterprise* in 1974 and developments of the F-14A initial service version should enter service in 1984. The improved F-14C has longer-life engines, a more capable radar and AIM-54C missiles with increased range and better guidance. More

dramatic improvements will come with the F-14D Super Tomcat incorporating new 29,000lbs thrust F101DFE engines, extra fuel, an even better radar and longer-range Phoenix missiles.

The F-14's partner in the Navy fighter squadrons is the McDonnell Douglas Phantom, currently serving in the F-4J, F-4N and F-4S versions. It is now an elderly design, having first flown in 1958, but it remains a highly-capable multi-role fighter. The F-4J succeeded the initial production F-4B, which had formed the backbone of the US Navy's fighter force for much of the 1960s. It has an improved fire control system, more powerful engines and greater fuel capacity. More than 200 F-4Bs were upgraded to this standard and redesignated F-4N. The F-4S has maneuvering slats on the outer wing to improve the aircraft's turn radius.

The Phantom's successor in the US Navy and US Marine Corps is the McDonnell Douglas F/A-18 Hornet, which will also replace the A-7 with Navy attack squadrons. It is a single-seater and is powered by two 16,000lbs GE F404-GE-400 turbofans. Maximum

speed is Mach 1.8 and combat radius in the fighter role is 425 nautical miles, increasing to 580nm for the attack mission. Wingspan is 37ft 6in, length 56ft and maximum takeoff weight 45,300lbs. Built-in armament is an M-61 Vulcan cannon and up to 19,000lbs of external ordnance can be lifted. For air-to-air combat two Sparrows are carried beneath the fuselage and two Sidewinders on the wingtips, with four wing hardpoints also available for AAMs.

The Hornet's principal sensor is the APG-65 radar, a multi-mode air-to-air and air-to-ground system. It can be simply operated by the pilot using throttle and control column mounted controls, with the data displayed on a HUD. Navigation, weapons control and sensor operation are all highly-computerized to ease the pilot's workload and to compensate for the lack of a second crewmember. For the ground attack mission the F/A-18 can carry a pod-mounted laser tracker and FLIR.

In comparison with the F-4, the Hornet is slower, but a Mach 2 air speed is reckoned to be unusable in air-to-air combat, and visibility from the Hornet's cockpit is much better. The avionics for the ground attack mission are equally as good as those of the A-7 and the Hornet has a much improved power-to-weight ratio which can be used to good effect in evading interception. A particular advantage of the F/A-18 is that it will replace two aircraft types, and so reduce the problem of spares holdings, which must perforce be limited by the carrier's relatively restricted storage space.

The carrier's offensive power for land or sea operations is provided by her attack squadrons. These are capable of using both conventional and nuclear weapons. The importance of the nuclear capability has declined since it was first introduced in the late 1950s; it remains a significant naval air mission, although conventional bombardment is today considered equally important. However a sustained campaign may quickly

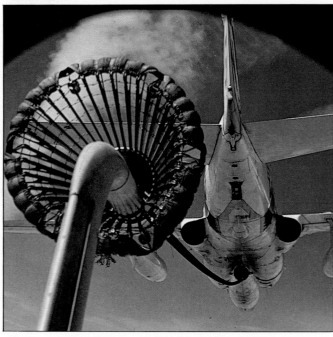

An A-6E Intruder (top) is prepared for a sortie from USS Coral Sea cruising in the East China Sea. An Intruder takes on fuel from a second A-6 fitted with a 'buddy' refueling pack (above). Two S-3A Vikings (right) of VS-31 The Topcats pose with arrester hooks extended. They belong to USS Dwight D Eisenhower's Air Wing Seven.

exhaust the ship's magazines and in a typical cycle of operations the ship will operate her attack squadrons for three days and withdraw on the fourth for replenishment. PGMs may reduce the expenditure of ordnance, but they are not suited to all types of target and their use may be restricted by poor visibility.

The Vought A-7E is at present the standard naval light attack aircraft and it equips 26 regular squadrons, with six reserve squadrons flying the older A-7B. In most essentials it is the same aircraft as the USAF's A-7D described in Chapter 3. The A-7E's highly-accurate navigation and bombing equipment gives it an all-weather capability which is further enhanced by the provision of FLIR pods for some aircraft. Weapons delivery is impressively accurate and a wide range of ordnance can be carried, including 'iron' bombs, anti-radiation missiles and PGMs. The A-7 also has a secondary fighter capability and can be armed with Sidewinder missiles to undertake combat air patrols. A similar emphasis on multi-role capability gives the F-4 and F-14 a secondary ground attack role.

The 14 medium attack squadrons fly the all-weather Grumman A-6E Intruder. This is a two seat aircraft designed to carry out an attack mission at low level, and if necessary in zero visibility, carrying either nuclear or conventional bombs. The A-6E is fitted with a high resolution APQ-148 multi-role radar operated by the bombardier/navigator, who is seated beside the pilot. It provides information for navigation, terrain avoidance,

and target acquisition and tracking. Navigation and weapons release data are supplied by an inertial navigation system and solid-state, digital computer. All-weather attack capability is further enhanced by TRAM (target recognition attack multisensor), a turret carrying forward-looking infra-red and laser equipment which provides the crew with a TV-type picture in most conditions of light and weather.

The A-6E is powered by two Pratt & Whitney J52-P-8 turbojets of 9300lbs thrust, giving a maximum speed of 557mph at 5000ft and a combat range of 1127 miles. Maximum takeoff weight is 58,600lbs for a catapult launch and dimensions include a span of 53ft and length of 54ft 9in. Ordnance load is up to 15,000lbs, which can be made up of bombs, CBUs, incendiary bombs, PGMs, ASMs, mines, rocket pods or nuclear weapons, with auxilary fuel tanks as a further option. Sidewinder AAMs can be carried for self-defense.

Each Intruder-equipped attack squadron includes a small number of KA-6D refueling tankers. These are A-6A aircraft modified by fitting a hose and reel installation under the fuselage. Four external fuel tanks are carried under the wing, with a fifth tank or 'buddy' refueling pack (to serve as a back-up if the main system fails) on the fuselage centerline. This gives a total capacity, including internal tankage, of 3844 US gallons, 3000 gallons of which is transferable. The KA-6D retains a limited attack capability in daylight conditions. This is another example of reducing the number of aircraft types aboard a carrier by modifying an attack aircraft for the refueling mission.

Carrier based ASW aircraft perform a vital protective mission by shielding the parent ship and its task force from underwater attack. Anti-submarine defenses are layered, with the fixed-wing Lockheed S-3A Viking providing an outer screen. The Sikorsky SH-3 helicopter maintains an inner line of defense in conjunction with escort ships, which themselves carry an ASW-capable helicopter. For many years the US Navy maintained a force of specialized ASW carriers, but these were retired in the early 1970s. Consequently the ASW squadrons have been integrated into the carrier air wings and are now embarked on all carriers.

The S-3A Viking, which became operational in 1975, carries a crew of four consisting of a captain, co-pilot, tactical co-ordinator and sensor operator. Power is provided by two 9280lbs thrust GE TF34-GE-2 turbofans, which are very economical consumers of fuel. Maximum range is 3368 nautical miles, with maximum speed 447 knots. Takeoff weight is 42,500lbs and dimensions include a span of 68ft 8in and length of 53ft 4in. An internal weapons bay can house nuclear or conventional depth bombs, homing torpedoes or mines (total internal weapons load is 2400lbs); rockets, mines, bombs or fuel tanks are carried underwing.

The Viking's specialized ASW equipment includes sonobuoys and a sonobuoy reference system, a radar, FLIR, magnetic anomaly detector (MAD) and elec-

tronic support measures (ESM) equipment to detect radar emissions. Data from these are stored and processed by a central computer and can be displayed to any crewmember. The aircraft also has a data link which can transfer information to its parent carrier and other ships and also receive information from them. However the Viking's systems are comprehensive enough for it to operate autonomously. Perhaps because so much complex equipment was packaged into a small airframe, the S-3A has had more than its share of unserviceability problems. It is hoped that this shortcoming will be rectified by the upgraded avionics of the S-3B.

Helicopter ASW squadrons are equipped with the Sikorsky SH-3 Sea King, which has been in US Navy service since 1961. It is powered by two GE T58-GE-10 turboshafts each of 1400shp. Maximum loaded weight is 20,500lbs and dimensions include an overall length of 72ft 8in and rotor diameter of 62ft. Maximum speed is 166mph, range is 625 miles and weapons load 840lbs. The principal ASW sensor is a 'dunking' sonar, which as the name suggests the helicopter lowers into the sea. The latest-standard SH-3H variant also has sonobuoys and an MAD. The standard offensive weapon is the homing torpedo.

Airborne early warning is the task of 15 squadrons flying the Grumman E-2 Hawkeye. The primary function of the AEW aircraft is to work in conjunction with fleet air defense fighters. It detects incoming targets at a range far greater than that possible with a ship-mounted radar (at 30,000ft the E-2C's radar can detect a bomber flying at the same altitude 450 miles away). Having detected inbound aircraft, and the Hawkeye can handle up to 30 contacts at once, the E-2 then directs fighters to intercept. The Hawkeye's radar can cover three million cubic miles of sky and its detection range can be increased by receiving radar data from an F-14 operating at the edge of this area. If an enemy aircraft is using its radar the emissions can be picked up by the Hawkeye at an even greater range. The E-2 is also useful as an airborne command post during offensive strike operations, for instance alerting the attack aircraft to enemy interceptors or directing their rendezvous with tanker aircraft. It can also be used for the detection of surface ships.

The E-2C is powered by two Allison T56-A-425 turboprops of 4508shp, giving a maximum speed of 348mph. Service ceiling is 30,800ft and maximum endurance is over nine hours, but a more typical sortie time is 3-4 hours. Maximum takeoff weight is 59,880lbs and dimensions include a length of 57ft 7in and span of 80ft 7in. For carrier stowage the wings can be folded (as with most shipboard aircraft) to reduce to 29ft 4in. More unusually the height of the E-2, with its radar mounted atop the fuselage, is a problem and to obtain clearance on the hangar deck the radome can be lowered by 1ft 8in, giving a height of 16ft 6in. The crew comprises two pilots, a radar operator, air control

operator and combat information center operator.

The E-2C's principal sensor is its APS-125 radar mounted in a rotating 24ft diameter radome above the fuselage. It is able to detect targets both over land and sea and can operate through ECM jamming with only a slightly reduced performance.

Electronic warfare is the province of the Grumman EA-6B Prowler, based on the A-6 Intruder but with the forward fuselage extended by 40in to accommodate two additional crewmembers. Ten US Navy squadrons currently fly the EA-6B, with the primary mission of escorting the carrier's strike force and jamming hostile radars during the time that the attack aircraft are over enemy territory. Its transmitters are powerful enough to allow it to carry out its mission from a 'stand-off' position outside the range of enemy SAMs. The aircraft's ECM capabilities are more fully discussed in Chapter 3.

Reconnaissance is another vital supporting role which the carrier air wing must undertake. Until its retirement from service in 1979, the US Navy had a highly capable reconnaissance platform in the RA-5C Vigilante. The RA-5C carried a wide range of equipment, including vertical, oblique and horizon-to-horizon cameras, flash pods for night photography, infra-red scanning and side-looking airborne radar. With the disappearance of the Vigilante only a single regular reconnaissance squadron was available for carrier deployment with the camera-equipped RF-8G Crusader, but this too was disbanded in 1982. However RF-4B Phantoms of the US Marine Corps can be de-

Electronic intelligence gathering is the role of the EP-3E Orion and EA-3B Skywarrior (above left). The E-2 Hawkeye (left) is an airborne early warning aircraft. The French Alizé (below) is the Aéronavale's shipboard ASW aircraft.

ployed aboard carriers, as can the more specialized EA-3B Skywarriors, which primarily undertake ELINT. The position has improved with the development of the TARPS for fitment to the F-14, some 50 of which are to be modified to use it. Each pod contains forward, oblique, vertical and panoramic cameras, plus an IR scanner. F-14s carrying TARPS can also be fully armed and so will not require a separate fighter escort.

An important support task is plane guard duty, which involves a Sikorsky SH-3G helicopter standing by to rescue any aircrews who have to ditch during carrier takeoff and landing operations.

Carrier flying is still a very exacting skill, in spite of modern approach aids, which can guide an aircraft into a fully automatic touchdown. The angled deck introduced on post-World War II carriers allows the pilot an unobstructed takeoff run should he miss the arrester wires on his first approach, but jet engines are relatively slow to respond to throttle movements and this can lead to landing accidents. Aircraft are generally launched by catapult and the modern carrier has four, allowing aircraft of up to 70,000lbs weight to be catapulted at 15 second intervals. The carrier's flight deck is generally crowded with aircraft, some with engines running, during the launching and recovery operations and the flight deck crew have a demanding and potentially dangerous job in handling, refueling and rearming them.

The French Navy operates two conventional carriers (*Clemenceau* and *Foch*) of limited size, with around 40 aircraft each. This is made up of two squadrons of Super Etendards operating in the attack and air defense roles, a squadron of Alizé ASW aircraft and two rescue and utility Alouette III helicopters. The Super Etendard is a Mach 1 multi-role naval fighter, armed with two 30mm cannon and AAMs or ASMs. The latter will include the ASMP tactical nuclear missile and AM-39 Exocet. An Argentine Super Etendard armed with Exocet sank the British destroyer HMS *Sheffield* and the merchant ship *Atlantic Conveyor* during the Falklands Crisis in 1982, although other attacks were defeated by British countermeasures, including the use of chaff rockets. The Alizé is a turboprop-powered three-seat ASW aircraft, fitted with radar and sonobuoys, which has a weapons load of 2000lbs and an endurance of up to 7½ hours. Two new nuclear-powered carriers are to be built as replacements for *Clemenceau* and *Foch*, with the first due to commission in 1991.

Because of the size and complexity of a super carrier, such ships are very expensive in terms of initial cost and manning. Therefore the V/STOL carrier would appear to offer a cheaper alternative. However at present the V/STOL carrier is not able to undertake the full range of naval aviation roles. The Royal Navy's *Invincible* Class carriers for example initially had no AEW capability until an AEW version of the Sea King helicopter was hastily developed in the aftermath of the Falklands Crisis. Although their BAe Sea Harriers are very versatile, inevitably such a small number of aircraft will be overwhelmed with demands for air defense, attack, reconnaissance and perhaps long-range ASW sorties. Therefore, far from being a substitute for the conventional carrier, the ASW carrier provides a limited measure of naval air power afloat for nations unable to afford a fully-capable carrier force. This can nevertheless be very valuable, as otherwise naval forces will have to rely on land-based air power for such missions as fleet air defense and long-range attack. Land based aircraft will perforce waste valuable time in transit to the naval forces' area of operations, they may not be on station when needed and indeed in an extreme case the warships may move out of range of the air support.

There will be three *Invincible* Class carriers, although the number in service with the RN may be reduced to two as an economy measure. They are equipped to operate Sea King ASW helicopters and Sea Harrier fighter-reconnaissance-strike aircraft, a typical air group being made up of five Sea Harriers and nine Sea Kings. However in an emergency the air group could be enlarged to more than 20 aircraft, by making use of a flight deck park as well as hangar storage. This capability was demonstrated during the Falklands Crisis, when *Invincible*'s air group was greatly expanded. The *Invincibles* and the ASW carrier *Hermes* are fitted with a ski-jump ramp, which in effect increases the Sea Harrier's takeoff run and enables a greater payload to be lifted.

The Sea Harrier FRS Mk 1 is based on the RAF's Harrier GR Mk 3, but is powered by the 21,500lbs thrust Rolls-Royce Pegasus Mk 104. The fuselage is redesigned to incorporate a nose-mounted Ferranti Blue Fox radar and the cockpit is raised improving the pilot's visibility. A new navigation system has been developed for over-sea flying and the Sea Harrier will be armed with the Sea Eagle anti-ship missile, as well as Sidewinder AAMs and air-to-ground ordnance. The Falklands Crisis has proved the Sea Harrier under combat conditions, although a war in European waters will be very different. Nevertheless the Fleet Air Arm's Sea Harriers performed brilliantly, shooting down 27 Argentine aircraft (including Mach 2 Mirage IIIs). Only seven Sea Harriers were lost, none of these in air combat. A total of 48 Sea Harriers has been ordered (14 since the Falklands Crisis) and there are three operational squadrons plus a training unit. The British version of the Sea King is a license-built version of the Sikorsky SH-3D, re-engined with 1500sph Rolls-Royce Gnome turboshafts. Most of the avionics, including Jezebel passive sonobuoys, is of British design and development by Westland Helicopters has proceeded independently of the US SH-3 variants. In 1982 a number were fitted with MAD gear of American origin. The Spanish navy has a small force of AV-8A Harriers and SH-3 Sea Kings which operate from the ageing carrier *Dédalo* (the former US wartime carrier

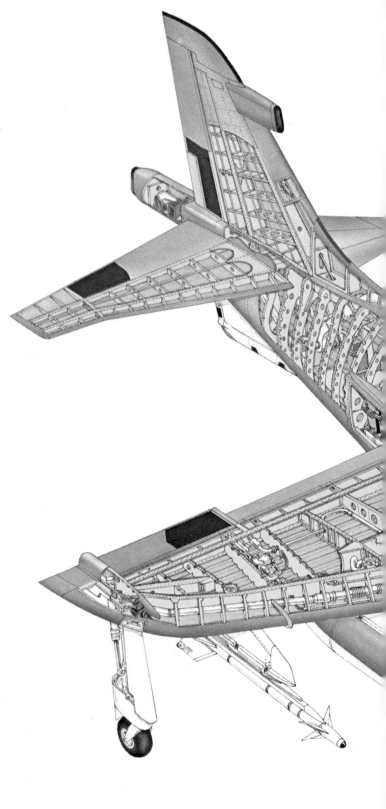

The BAe Sea Harrier FRS Mk 1 proved its worth during the Falklands campaign in 1982, flying from Invincible class carriers and HMS Hermes (cutaway artwork). This Fleet Air Arm Sea Harrier (top) carries Sidewinder air-to-air missiles.

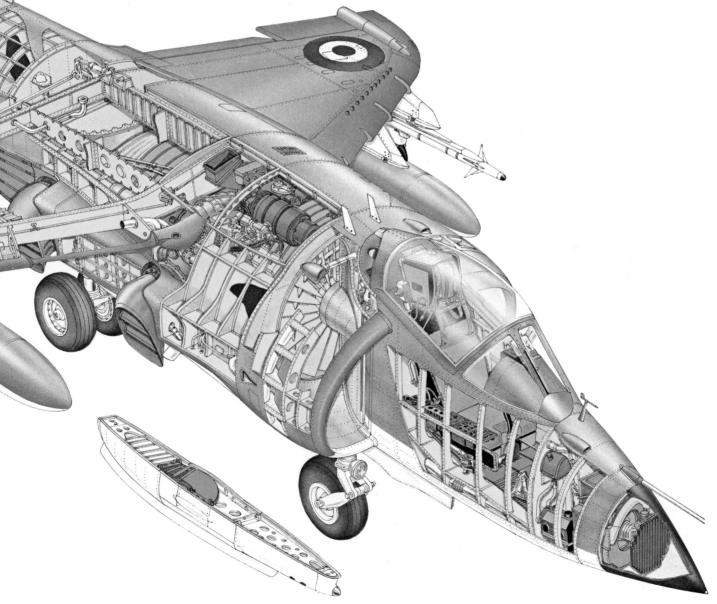

The Soviet helicopter cruiser Moskva (above) carries Ka-25 Hormones (right). The British Lynx (below right) is armed with Sea Skua missiles. The US Navy SH-2F (top right) trails a MAD drogue. An ASW torpedo is carried beneath the sonobuoy launchers.

Cabot, rebuilt for ASW work in the 1950s). They will in due course transfer to the new sea control ship *Principe de Asturias*, launched at Ferrol in 1982.

The Soviet Union's counterparts to the *Invincible* are the *Kiev* Class ships, each of which carries about 12 Yak-36 Forger VTOL combat aircraft and 24 Kamov Ka-25 Hormone ASW and missile guidance helicopters. The first two ships of the class are now in service, with two more under construction. The Yak-36MP Forger like the Sea Harrier, can undertake a number of roles, including fleet air defense, reconnaissance and attack. However, it is a less capable aircraft than the Sea Harrier, as it only has a small ranging radar and an IR sensor, which are of limited use for interception and even less for reconnaissance.

The major difference between the Sea Harrier and Yak-36 is that the latter is powered by a main turbojet with vectored nozzles and two separate lift-jets, rather than the British fighter's single vectored-thrust powerplant. The Soviet system is far less flexible and restricts the Forger to vertical takeoff operations – with a consequent limiting of payloads – as STOL would present too many stability problems. Forger spans 24ft 6in and length is 52ft 6in, with maximum takeoff weight

around 28,600lbs. Level speed is around Mach 1.05 and mission radius for high altitude reconnaissance up to 300 miles.

The introduction of V/STOL may lead to the use of merchant vessels as auxiliary aircraft carriers. The US Navy and the Royal Navy are investigating the possibility of converting large container ships into V/STOL carriers by installing a prefabricated runway and limited aviation support facilities (US Project Arapaho). During the Falklands Crisis the British took this a step further by fitting two large container ships with aircraft facilities. Looking much further into the future, the US Navy is interested in a Mach 2 VTOL fighter for service at the end of the present century.

If few warships can undertake the whole range of naval air operations, conversely there are few that do not have a measure of air capability by virtue of the ubiquitous helicopter. In the US Navy frigates, destroyers and guided missile cruisers are all equipped to operate the Kaman SH-2F Seasprite helicopter. More than 100 of these multi-purpose helicopters have been procured. They have the primary task of ASW and are able to operate at distances of up to 80 nautical miles from the parent ship.

The Seasprite is equipped with a search radar, MAD and sonobuoys. A data link relays information to the parent ship where most of the processing is carried out. With a crew of only three (two pilots and a systems operator) the Seasprite cannot accommodate the equipment and personnel necessary for completely auton-

omous ASW operations. Armament comprises two Mk 46 anti-submarine torpedoes. The Seasprite's secondary roles include anti-ship missile defense, using ECM jamming transmitters to interfere with an incoming missile's radar guidance. Rescue tasks are also undertaken by the Seasprite.

A new shipboard helicopter, the Sikorsky SH-60B SeaHawk is under development and first deliveries are due in late 1983. It is a development of the US Army's UH-60A Black Hawk. Modifications for shipboard operation include a folding main rotor and tail boom to facilitate stowage, salt water corrosion proofing, and a haul-down system to assist landing on a warship's pitching helicopter pad. ASW patrols can be undertaken 100 nautical miles from the parent ship, using a similar range of equipment to that of the Seasprite. The SeaHawk has also been designed to provide over-the-horizon targeting information for ship or submarine-launched Harpoon anti-ship missiles. The current requirement is for 204 UH-60Bs and a modified version with dunking sonar will replace SH-3 Sea Kings aboard the carriers.

Six of the European NATO navies fly the Anglo-French Lynx multi-purpose helicopter from frigates and destroyers. In British service the Lynx's primary role is to deliver ASW weapons, with a secondary role against surface ships, using its Ferranti Seaspray radar to search for hostile vessels. It can then attack with its own Sea Skua missiles or provide target information to the parent ship. Anti-submarine armament can be two

An SH-3 Sea King (above) operates its dunking sonar. Soviet Yak-36s operate from the Minsk (inset right). AV-8A Harriers (right) fly with VMA-231. US Marine helicopters (bottom) include the Sidewinder-armed AH-1T (left), CH-53A (center), here used for minesweeping with the US Navy's HM-12 and UH-1E (foreground) and AH-1J.

homing torpedoes or two depth charges. The French Navy places greater emphasis on ASW capability and its Lynxes are fitted with a dunking sonar. Lynx is powered by two 900shp Rolls-Royce Gem Mk 100 turboshafts, giving a maximum speed of 184mph. Maximum takeoff weight is 8000lbs and dimensions include a fuselage length of 38ft 4in and rotor diameter of 42ft.

The Soviet Union's standard shipboard helicopter is the Kamov Ka-25 Hormone, which is carried aboard frigates, destroyers and cruisers. The battlecruiser *Kirov* can accommodate up to five Ka-25s, while the helicopter carriers *Moskva* and *Leningrad* operate 36. The helicopter is produced in three versions. Hormone-A is an ASW helicopter while Hormone-B provides mid-course guidance for surface-launched anti-ship missiles, such as the 340-mile-range SS-N-12. Hormone-C is a utility and search and rescue variant.

The Ka-25 is powered by two 990shp Glushenkov GTD-3 turboshafts driving two, co-axial, contra-rotating rotors. Maximum speed is about 130mph and range on internal fuel is 250 miles, which can be increased to 400 miles when auxiliary fuel tanks are carried. Maximum takeoff weight is 16,500lbs, overall length is 32ft and rotor diameter 51ft 8in. The crew comprises two pilots and two or three systems operators. ASW equipment includes a search radar, MAD gear and dunking sonar. A weapons bay beneath the cabin can carry homing torpedoes, or depth charges.

A new Soviet naval helicopter, which may eventually replace Hormone, was first observed in October 1981 aboard the guided-missile destroyer *Udaloy*. Code-named Helix, the helicopter makes use of the same rotor system as the Ka-25, but the powerplant is probably new and the fuselage is appreciably bigger. This suggests that improved and more comprehensive ASW equipment is carried and Helix certainly has a better

endurance than the Ka-25. It could also provide the basis for an amphibious assault helicopter, with the capacity for up to 20 troops in the cabin.

The United States is the only nation with a significant capability to conduct amphibious operations worldwide, although the Soviet Union has less well developed amphibious forces. Such operations are the major role of the US Marine Corps, which has its own air arm. Marine aviation units can meet all the air support requirements of a landing force, including close air support, battlefield and deeper interdiction, counter-air operations, beachhead air defense and tactical reconnaissance. Helicopter units can airlift part of the landing force during assault or evacuation and can also supply units ashore with ammunition and other vital supplies. The Marine Corps also has a small force of tactical fixed-wing transports, which have the secondary task of air refueling.

The fixed-wing combat aircraft of the US Marine Corps can operate either from shore bases or from the US Navy's carriers, the transition from shipboard to land being made as early in the assault as practicable. Hence the Marine Corps interest in the V/STOL Harrier, with its minimal airfield requirements. Three light attack squadrons currently fly the AV-8A and a further five are equipped with the McDonnell Douglas A-4M Skyhawk fighter bomber. From 1985 all eight light attack squadrons will begin to convert onto the AV-8B Harrier II. Five medium attack units are equipped with the Grumman A-6E Intruder. In general the Marine Corps flies the current US Navy type appropriate to its role. However, there are exceptions to this rule, notably in the case of the V/STOL Harrier, to which the Marines are firmly committed while the Navy adopts a more cautious attitude to V/STOL at sea.

Another instance of the divergence of Navy and Marine Corps procurement policies is provided by the A-4 Skyhawk. During the 1960s the Skyhawk was the Navy's standard light attack aircraft, but was superseded by the Vought A-7 and retired from first-line Navy service in 1975. However the Marine Corps elected to retain the Skyhawk, rather than re-equip with the heavier and more costly A-7, which offered few advantages in the CAS role, notwithstanding its overall improvements in all-weather attack capability.

The current Marine version of the Skyhawk is the A-4M. This is powered by a 11,200lbs thrust Pratt & Whitney J52-P-408A turbojet, which gives a maximum speed of Mach 0.88 at sea level. The takeoff run is 2730ft at a weight of 23,000lbs and combat radius is 385 miles. The takeoff run represents a 20 percent improvement over earlier A-4s and the rate of climb (4800ft per minute from sea level) is also considerably greater. Normal takeoff weight is 24,500lbs and span is 27ft 6in with overall length 40ft 4in. Two 20mm cannon are mounted in the wing roots and have an ammunition capacity of 200 rounds each. A 6000lbs ordnance load is typical, although this can be raised to more than

A petty officer (above) loads sonobuoys onto an SH-2 helicopter. A P-3C Orion (above right) of VP-24 is pictured on patrol. A P-3's radar display is shown (right).

9000lbs. The A-4M is fitted with an angle-rate bombing system and HUD for the pilot.

Twelve Marine fighter/attack squadrons fly the F-4J, N and S versions of the Phantom. The primary mission of these units is to gain air superiority over the beachheads, but they are also equipped and trained to undertake CAS. The Phantoms will be replaced by F/A-18 Hornets from 1982, with the first squadron becoming operational in 1983. Electronic warfare missions are carried out by 15 EA-6B Prowlers and reconnaissance is provided by RF-4B Phantoms, equipped with cameras, infra-red and side-looking radar.

Three observation squadrons fly the OV-10A Bronco for battlefield surveillance and FAC, with secondary light attack and troop transport roles (five paratroopers can be carried in some discomfort). For night observation and attack missions a number of Broncos have been converted to OV-10Ds by fitting FLIR and laser sensors. The fixed-wing inventory is completed by three squadrons equipped with KC-130F and KC-130R variants of the Hercules, which undertake air refueling and tactical transport missions. In the tanker role a 3600 gallon tank is carried in the cargo compartment and refueling is by means of pod-mounted hose and drogue units carried underwing.

The helicopters supporting an amphibious assault are carried by amphibious assault vessels of the *Iwo*

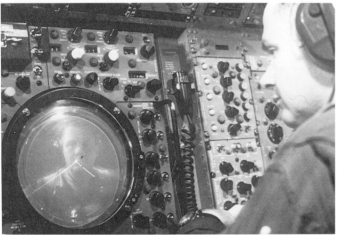

Jima Class (seven in service) and *Tarawa* Class (five ships). Each can accommodate 30-35 helicopters and is also able to operate Harriers. The helicopter complement is generally a mixture of heavy, medium and light transport machines and a number of attack helicopters. Six squadrons are equipped with the heavy Sikorsky CH-53D/E Sea Stallion, nine with the medium Boeing-Vertol CH-46D/E/F Sea Knight, six with the Bell UH-1E/N light helicopter and three with AH-1J/T Sea Cobras for attack.

The CH-53D is powered by two GE T-64-GE-413 turboshafts of 3230shp each. Takeoff weight is 36,628lbs and dimensions include a rotor diameter of 72ft 3in and length with rotor and tail pylon folded of 56ft 6in. Speed at sea level is 166 knots and combat radius is 100 nautical miles. There is accommodation for 38 troops and up to 8000lbs of cargo under normal operating conditions, although under overload conditions 12,740lbs can be lifted. The CH-53 can carry underslung loads, including artillery weapons, and has been used to recover shot-down aircraft. The more powerful CH-53E (with three 3695shp T64-GE-415 turboshafts) entered service in 1981. Its lifting capacity for external loads is double that of the CH-53D.

The medium lift CH-46D/E/F can carry 26 troops, plus its crew of three. It is powered by two 1400shp GE T-58-GE-10 turboshafts, giving a maximum speed of 166mph and a 250 mile range. Maximum weight is 23,000lbs and overall length is 84ft 4in. The twin tandem rotors are each of 51ft diameter. As the Sea Knight first became operational with the Marine Corps in 1964 its replacement would be due on grounds of age alone. The requirement has become more urgent because there are insufficient medium-lift helicopters in service to meet current needs. In the short term the Marines will make an off the shelf purchase to increase the medium helicopter inventory. By the early 1990s the joint service JVX requirement is due to produce a new medium assault machine and this is likely to be a tilt-wing aircraft rather than a helicopter.

The UH-1 squadrons are responsible for airborne command and control during the helicopter assault and subsequent ground fighting, but they do have a secondary transport mission and up to eight troops can be carried. Their aircraft is basically similar to the US Army's UH-1, fully described in Chapter 8. The attack squadrons' Bell AH-1 is also derived from an Army helicopter. However the AH-1J Sea Cobra differs from its single-engined Army counterpart in having twin 1800shp UAC T-400-CP-400 turboshafts. Speed is 207mph and range 360 miles. The armament comprises a built-in 20mm cannon, with rocket or machine gun pods carried on stub wings. The AH-1T has uprated engines and carries TOW antitank missiles. It is also to be equipped with Sidewinder AAMs to counter Soviet Hind-D helicopter gunships and Hellfire advanced 'fire and forget' antitank missiles.

Other NATO amphibious warfare forces are on an altogether smaller scale than the US Marines. However the British have a Royal Marine Commando Brigade trained in the assault role to reinforce NATO flank areas and two specialized amphibious assault ships, *Intrepid* and *Fearless*, which could be augmented by the ASW carrier *Hermes* or an *Invincible* Class ship. Westland Wessex HU5 and Sea King HC4 assault transport helicopters equip two Fleet Air Arm squadrons (Nos 845 and 846). The Wessex can carry 16 troops, while the Sea King has room for 28. Furthermore the Sea King can lift an underslung load of 8000lbs, twice that of the older Wessex. The Royal Marine Commandos have their own air units, which fly Gazelle AH1 helicopters for scouting and Lynx AH1 antitank helicopters. The Soviet Union's amphibious capabilities are equally short range and will probably concentrate on seaward flanking attacks in support of a land advance, for example in Scandinavia. An especially important Soviet objective is likely to be the capture of the NATO-controlled exits to the Baltic and Black Sea. Naval Aviation flies the Su-17 Fitter-C which could support such operations, as well as undertaking short range anti-shipping strikes.

The battle against the submarine has been a central theme of warfare in the 20th century and in both world wars these vessels came close to deciding the outcome of the conflict. Yet the significance of anti-submarine warfare (ASW) in World War III will be immeasurably greater. This is because the application of nuclear propulsion to underwater craft has produced the true submarine, which is capable of remaining continuously submerged for periods of two months or more. Furthermore, when these potent new craft were armed with sea-launched ballistic missiles carrying thermonuclear warheads, ASW became quite literally of vital concern.

The aircraft is an important participant in ASW operations, although surface ships and submarines are also capable of detecting and killing enemy submarines. In practice the three systems will complement each other, with the nuclear-powered hunter-killer or attack submarine able to operate close to enemy naval bases, where surface ships or ASW aircraft would be extremely vulnerable. The anti-submarine warship has a far greater endurance and weapons' capacity than the aircraft, while the latter is quick to react to a submarine contact and can switch its area of operations at short notice.

The main problem in ASW is locating the target, as once a submarine's position has been pinpointed it is vulnerable to a wide range of weapons. The most effective of these is the acoustic homing torpedo, although depth charges can also be useful. The effects of explosives detonated under water are more powerful than on land, due to water's incompressibility, and if a nuclear depth charge is employed then the kill becomes even more certain.

Sonar is still the most effective method of anti-submarine detection, but it has definite limitations due to short range and erratic performance. Therefore other detection devices are carried by ASW aircraft to increase the chances of success. Radar is useful in detecting a surfaced submarine, or one with a periscope or schnorkel above water. Not all modern submarines are nuclear-powered and diesel-electric boats must recharge their batteries for about 20 minutes in every 24 hours, running at a shallow depth with a schnorkel tube (or snort) above water. It is also possible to detect the diesel fumes from a snorting submarine, although the Autolycus equipment developed for this by the RAF has now gone out of favor. Lastly any submarine which transmits a radio message or uses radar can be detected by ESM equipment.

It is not only the aircraft's sonar which helps in the submarine hunt. Barriers of fixed sonars can be laid in areas where enemy submarines are expected to operate. There is one off the US Atlantic coast and another guards the Greenland-Iceland-UK gap through which Soviet submarines based on the Kola peninsula must pass to reach the Atlantic. These sonars relay information to ground stations, which can in turn alert patroling aircraft.

ASW aircraft carry sonobuoys, which are dropped in patterns and transmit data back to the aircraft. They are of two types. Passive sonobuoys pick up underwater

An RAF Nimrod MR Mk 2 maritime patrol aircraft (top) circles its base at Kinloss in Scotland. The Soviet Il-38 May ASW patrol aircraft is pictured in flight over the Indian Ocean.

sounds, enabling the source to be identified and a bearing taken on it. A computer aboard the ASW aircraft compares the sonobuoy's signal with a stored memory. It can not only distinguish a submarine from other sources of sound, such as marine life, but can precisely identify the nationality and type of boat. Passive sonobuoys are useful in the early stages of a submarine hunt. As they do not transmit sound signals, they cannot alert a submarine to the presence of a hunting aircraft. Yet since they are laid in fields, the data from two or more sonobuoys will provide the position of the submarine.

Once a submarine has been located, more precise information is needed to allow an attack to be set up. This is provided by active sonobuoys, in effect miniature ships' sonars, which transmit a sound pulse through the water. When this bounces back off the submarine's hull, it will give range, bearing and speed to the sonar operator. However it will also alert the enemy crew and thereafter the attack must be made swiftly to prevent the submarine from slipping away. The weapon's release run may be assisted by an MAD, which will detect the local variation in the earth's magnetic field caused by the metal hull of the submarine. However, the disadvantages of the MAD are its short range and the fact that it will give much the same reading for a submerged wreck as an undamaged submarine.

The Soviet Union maintains more than 300 nuclear and diesel-electric submarines and these are the primary target for NATO ASW forces. The US Navy's standard shore-based anti-submarine patrol aircraft is the Lockheed P-3 Orion, with P-3Bs and P-3Cs equipping 24 front-line squadrons and older P-3As and P-3Bs, serving with reserve squadrons. The Atlantic Fleet patrol wings are based at Brunswick, Maine and Jacksonville, Florida, with detachments operating from Iceland, Sicily, the Azores and Bermuda. Pacific operations are even more far-flung, with squadrons flying from Moffett Field, California, Barbers Point, Hawaii, Adak in the Aleutians and Misawa, Japan – and even this listing is not complete.

The latest model of the Orion, the P-3C remains in production, with just over 200 aircraft delivered by early 1982 against an eventual requirement for 275. A conversion of the Electra civil airliner, the Orion has ample accommodation for ASW equipment and operators, the total crew being 10-12. The P-3C is powered by four 4910shp Allison T-56-A-14 turboprops, which give a top speed of 473mph and an endurance of over 17 hours. Maximum takeoff weight is 135,000lbs and dimensions include a span of 99ft 8in and an overall length of 116ft 10in. ASW sensors include radar, ESM, MAD, active and passive sonobuoys and the associated acoustic processing equipment, low-light TV or FLIR. This equipment is being continually modified and improved; the Update III program will introduce a new acoustic processor and radar from 1984 onwards. Offensive armament includes up to 15,000lbs of depth charges, homing torpedoes or mines in an internal bay and on 10 external hardpoints. Unguided rockets or AGM-12 Bullpup ASMs can be carried under the wings and P-3C Update II aircraft can carry the AGM-84A Harpoon anti-ship missile.

The British Aerospace Nimrod, which equips four RAF squadrons, was the first jet powered ASW aircraft anywhere in the world. The Nimrod, like the P-3, was based on a commercial transport aircraft, in this case the Comet. Power is provided by four 12,160lbs thrust Rolls-Royce Spey Mk 250 turbofans, which give a maximum speed of 575mph and an endurance of 12 hours. Until the economical turbofan was developed, jet engines were considered too 'thirsty' for use on maritime patrol aircraft. For economical cruise on patrol, two of the Speys are shut down. However their reserves of power are available once contact with a submarine has been made and they allow a rapid flight out to the patrol area.

The Nimrod spans 114ft and is 126ft 9in in length, with a maximum takeoff weight of 177,500lbs. The normal crew complement is 12. ASW equipment includes radar, active and passive sonobuoys, MAD, ESM and a 70 million candle power searchlight in the starboard wing-mounted fuel tank. An underfuselage weapons bay houses a variety of armament, including the Sting Ray lightweight homing torpedo. The Nimrod fleet is being progressively updated to MR Mk 2 standard, with a new computer-assisted EMI Searchwater radar with multiple target tracking ability and improved range and resolution. Improved acoustic processing is incorporated and a new ESM system is under development.

Another important NATO ASW patrol aircraft is the Dassault-Breguet Atlantic, which is operated by France, West Germany, Italy and the Netherlands. It is a twin-engined aircraft, powered by 5190shp Rolls-Royce Tyne 21 turboprops, and carries a crew of 12. Maximum endurance is 18 hours, but a more typical patrol would last for 11 hours. France is to re-equip the four Aéronavale flottilles currently flying this type with the Atlantic Nouvelle Generation, which is a basically similar airframe with completely updated avionics.

The Soviet Union's commitment to the ASW mission is at least as strong as that of NATO. Indeed the threat from missile-armed submarines may have been the primary reason for the Soviet fleet expanding its area of operations into the North Atlantic and Mediterranean. The standard Soviet long-range maritime patrol and ASW aircraft is the Ilyushin Il-38 May, which is based on the Il-18 airliner. It is powered by four 5200shp Ivchenko AI-20M turboprops, which give a maximum speed of about 400mph. Range is over 5000 miles and endurance 16 hours. The Il-38's ASW sensors include radar, MAD, sonobuoy equipment and perhaps ESM. However, the performance of these systems is unlikely to be up to Western standards, not only because of the Soviet Union's lack of expertise in

electronics and computer technology, but also because it lacks the West's long experience of airborne ASW operations.

Shorter range ASW patrols are undertaken – most unusually – by the Beriev Be-12 Mail amphibian flying boat. Most major nations phased out flying boats in the 1950s in favor of land-based aircraft, which have better payload/range characteristics. The only other exception is Japan, which operates the four-turboprop Shin Meiwa PS-1 flying boat. The Be-12 is powered by two 4190shp Ivchenko AI-20D turboprops, which give a maximum speed of 380mph and a range of 2500 miles. A search radar is carried in the nose and an MAD is mounted at the end of a boom in the tail. This is a common practice for MAD installations, as it removes the sensor from sources of metal in the aircraft. An alternative arrangement, often used by helicopters, is to attach the MAD to a drogue which is towed behind the aircraft. Sonobuoys may be carried by the Be-12 and a flying boat can of course alight on smoother water to use a hull-mounted sonar.

Another short-range Soviet ASW aircraft is the Mil Mi-14 Haze, which is an adaptation of the Mi-8 transport helicopter. The Mi-14 is an amphibian, with a boat hull and land undercarriage, and is operated from land bases. ASW equipment includes a search radar, towed MAD and possibly dunking sonar, while armament is likely to be homing torpedoes or depth charges.

Long range ASW aircraft also have a maritime reconnaissance role, which involves surface surveillance rather than submarine hunting. This mission can also be undertaken by AWACS and AEW aircraft such as the Boeing E-3A Sentry, Nimrod AEW Mk 3 and Grumman E-2 Hawkeye. The USAF also uses the B-52 on such missions, flying at high altitude for maximum radar coverage and descending to identify suspicious contacts visually. Tu-20 Bears of Soviet Naval Aviation are operated in a similar way, making use of bases in Cuba, Angola, South Yemen, Ethiopia, Syria, Libya and Vietnam. Satellite surveillance is also useful for the maritime mission. Offshore oil and gas installations have further complicated the ocean surveillance task, the RAF for example flying special 'Offshore Tapestry' sorties with Nimrods to safeguard these vulnerable assets.

Anti-shipping strikes are the province of naval attack aircraft, such as the US Navy's A-7E and A-6E, the French Super Etendard and British Buccaneer. The ASM has replaced the air-launched torpedo, which made the anti-shipping aircraft of World War II so vulnerable to defensive fire. However free-fall bombs and unguided rockets are still used on these sorties.

The US Navy's AGM-84A Harpoon anti-ship missile, which has a stand-off range of some 70 miles, arms A-6Es and P-3C Orions. France's AM-39 Exocet, which received a lot of publicity because it sank the British destroyer *Sheffield* during the Falklands crisis, is less capable with a range of around 40 miles. These

The Atlantic NG (above left) is to be the Aéronavale's new patrol aircraft. A CH-53A (above) tows a minesweeping sled during the clearance of Haiphong harbor.

missiles, and the British Sea Eagle which is entering service, make use of inertial guidance for most of their flight, with active radar homing for the final attack phase. Because of the great importance attached to defending the Baltic exits from a Soviet breakout, the West German Navy will equip two wings with the Tornado for anti-shipping strikes in this area. They will be armed with 20-mile range Kormoran missiles and the Italian air force Tornados will also use this weapon. A general weakness of stand-off anti-shipping missiles is their susceptibility to ECM jamming. The carrier aircraft must obtain the target's position on its radar to program the missile's inertial guidance system before launch and this can alert the ship under attack. Hence the need for unguided bombs and rockets as a back-up to the ASMs.

Soviet anti-shipping missions are generally flown by long range bomber aircraft, such as the Tu-26 Backfire, Tu-22 Blinder and Tu-16 Badger. They carry various ASMs with maximum ranges of between 50 and 180 miles, which can have either nuclear or conventional warheads. The US Navy's aircraft carriers are of course one of their prime targets. Shorter range anti-shipping missions are flown by the Su-17 Fitter or by shipborne Yak-36MP Forgers.

Sea mining remains an important way of attacking enemy naval and merchant shipping and many aircraft are capable of aerial minelaying. Carrier attack aircraft including the A-7E and A-6E can lift small loads of mines and will be used to mine heavily defended areas, such as enemy harbors. The A-6 Intruder was responsible for the mining campaign against North Vietnamese harbors in 1972. Rather heavier loads are carried by S-3 and P-3 aircraft.

Mines can be sown in defensive barriers and for this task an aircraft with a large payload is required. The Lockheed C-130 and other transports have been con-

sidered for this task and the B-52D is able to lift more than eighty 500lb mines. A more sophisticated mine, the US Mk 60 Captor, is in effect a moored acoustic homing torpedo which can distinguish friendly from hostile targets. Such weapons will be of especial value in the choke points used by submarines in transit to the operational areas. The Soviet Union, with its limited access to the oceans, is especially vulnerable to the effects of mine warfare. Perhaps for this reason, it is well aware of the value of mine warfare and has the largest stockpile of naval mines in the world. All Soviet long range naval aircraft are capable of minelaying.

Mine clearance is of course of great concern to all navies and the US Navy makes use of specially equipped helicopters to supplement the conventional mine hunter and minesweeping ships. At present these are RH-53D variants of the Sikorsky Sea Stallion, which tow a hydrofoil sled containing the mine-sweeping equipment along the sea surface. Their replacement is the three-engined MH-53 E Super Stallion which can tow a larger sled and operate in heavier seas. The great advantage of a helicopter minesweeper is its immunity from the effects of an underwater explosion. RH-53Ds are detached to vessels of the Atlantic and Pacific Fleets, generally operating from amphibious warfare ships. They can also be airlifted in the USAF's C-5A Galaxy transports. The helicopter mine counter-measures concept has been proved during the clearance of Haiphong Harbor in 1973 and of the Suez Canal a year later after the 1973 Yom Kippur War.

Mirage 2000 prototypes escort a Mirage 4000. The Mirage 2000 is due to enter service with the Armée de l'Air in 1983.

5. AIR DEFENSE

Air Defense

The defense of friendly air space from intruders is a vital task alike in peace as in war. Air defense units maintain a constant alert 24 hours a day, 365 days a year. They are therefore better prepared than most elements of the armed forces to switch from a peacetime footing to all-out war. However this advantage is counterbalanced in the West by the serious neglect of air defense forces, which have for years operated elderly aircraft and missiles. Not so in the Soviet Union, where air defense – or more properly aerospace defense – enjoys a high priority in manning and equipment.

Air defense in essence is the safeguarding of national air space. All air traffic is identified and tracked. Commercial aircraft must file a flight plan before takeoff and any deviation from this or wandering from clearly defined airways corridors can lead to the errant aircraft being intercepted and investigated. Normally such aircraft fly in controlled air space under the direction of ground controllers, so keeping track of thousands of commercial aircraft is not such a daunting task as at first may be supposed. Military aircraft too are subject to ground control and the civil and military air traffic control organizations work closely together. Any unidentified aircraft approaching a national air defense identification zone (ADIZ) will be intercepted, identified and, if necessary, destroyed. The air defenses will be very active in the period of tension preceding an outbreak of war and once hostilities begin they have the vital role of protecting the homeland against strategic attack and the tactical forces' lines of communication from disruption by far-ranging interdictor aircraft.

Ground based and airborne elements of an air defense system work very closely together and therefore C^3 (command, control and communications) are of the greatest importance. Normally ground radars provide the first warning of attack, although with the increasing deployment of airborne early warning (AEW) and control aircraft this emphasis is shifting. Once a threat has been recognized interceptors can be scrambled to deal with it. An alternative which results in an even quicker reaction time is to have the interceptors already airborne on combat air patrol (CAP). This expedient can be valuable if an attack is anticipated with some degree of certainty, but as a normal operational procedure it is very wasteful.

Interceptors are directed by a ground or airborne controller until they can pick up their target on the aircraft's radar. Normally they must positively identify the target as hostile before engaging it with missiles or guns. Although aircraft, including cruise missiles, are the primary targets of air defense forces, their responsibilities also extend into space. Ballistic missile early warning is now a routine activity and great efforts are being made by both the United States and the Soviet Union to develop reliable anti-ballistic missile (ABM) weapons.

The responsibility for the defense of the air space of

The newest USAF interceptor aircraft is the F-15 Eagle, an aircraft of the 1st TFW is shown (top). Canada assigns four squadrons of CF-101F Voodoos (center) to NORAD. The F-106A (center right) is still an important USAF interceptor. The aircraft are backed by ground-based radars (top right). Two squadrons of Lightning interceptors contribute to the air defense of the United Kingdom (above).

the United States and Canada is entrusted to the North American Aerospace Defense Command (NORAD). This arrangement between the Canadian and US governments, integrates early warning, C³ and interceptor forces for space and atmospheric defense into a single organization. Nerve center of the command is the underground operations complex in Cheyenne Mountain, Colorado, where a computerized system manages all space surveillance, missile launch warning and aircraft early-warning systems.

Unfortunately, these sophisticated systems, often termed the air defense ground environment (ADGE) are not complemented by an up-to-date interceptor force. Most current USAF interceptor aircraft are 20 years old, with weapons and radar of limited use against low-flying aircraft. These shortcomings have been recognized and a modernization program is underway to equip the regular USAF fighter interceptor squadrons (FIS) with the F-15 Eagle and assign E-3A AWACS (airborne warning and control system) aircraft to North American air defense.

During the 1960s and 70s the Soviet manned

bomber threat was, quite correctly, considered to be less significant than that from missiles. Consequently the USAF's air defense force was run down from a peak strength of over 40 regular squadrons in the late 1950s to its present level of six regular squadrons, backed by part-time ANG units. However, as the Soviet strategic bomber force has not disappeared and indeed may shortly introduce a new and effective inter-continental range manned bomber, the cutbacks have probably gone too far. In 1982 regular and ANG squadrons operated 153 F-106s, 90 F-4s, 36 F-101s and 18 F-15s. The F-101 units of the ANG are shortly to be re-equipped with F-4s.

In 1979 the USAF phased out its Aerospace Defense Command (ADCOM) as a separate operational headquarters, although the function and authority of NORAD's Cheyenne Mountain headquarters (the Aerospace Defense Center) remained unchanged. ADCOM's radars, control centers and interceptors are now managed by Tactical Air Command, while SAC is responsible for missile warning and space surveillance sensors. TAC's air defense interceptor units comprise:

Unit	Base	Equipment
5th FIS	Minot AFB, ND	F-106A
48th FIS	Langley AFB, Va	F-15C
49th FIS	Griffiss AFB, NY	F-106A
57th FIS	Keflavik AB, Iceland	F-4E
87th FIS	Sawyer AFB, Mi	F-106A
318th FIS	McChord AFB, Wa	F-106A

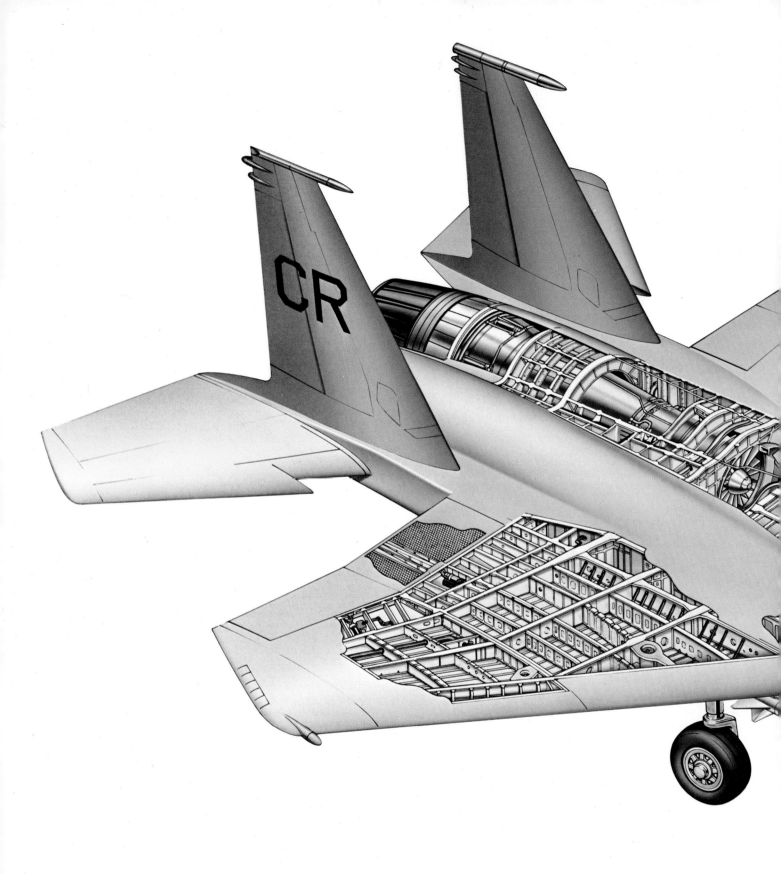

F-15 Eagles will replace F-106s as the USAF's standard interceptor (cutaway artwork).

The Air National Guard provides a further 10 squadrons, four with F-106s, four with F-4s and two with F-101s. In addition the Canadian Armed Forces assign three squadrons of CF-101B Voodoos to NORAD and these are to be re-equipped with more modern CF-18 Hornets.

The Western European members of NATO do not have so clearcut an air defense problem as that of the North American partners. West Germany in particular is likely to find that her territory becomes the battleground in the early stages of World War III. Therefore her air defense is likely to merge with the air superiority mission. There is, however, a real distinction between the two roles. The air superiority fighter will engage high performance aircraft of its own kind over a confused battlefront. In contrast the air defense interceptor will operate over friendly territory (perhaps an inappropriate term for the North Sea or Canadian Arctic), under the control of ground radar against relatively cumbersome bomber and interdictor aircraft. However, other European NATO powers, in particular the United Kingdom, France and Spain, will face a threat that can best be countered by specialized interceptors.

The Soviet MiG-25 Foxbat interceptor (top) has an outstanding speed and climb performance. In order to improve the United Kingdom's air defenses, the RAF plans to equip its Hawk weapons trainers with Sidewinder AAMs (center) to provide air cover over such vital targets as airfields. The F-4 (above) serves with Air National Guard interceptor squadrons.

The West German Luftwaffe operates two wings (JG 71 and JG 74) of F-4F Phantoms in the air superiority/air defense role. These are bolstered by two squadrons of Phantom FGR Mk 2s with RAF Germany and USAFE F-4E and F-16A units. The United Kingdom's air defense interceptor force comprises five squadrons of Phantom FG Mk 1s and FGR Mk 2s, plus two squadrons of shorter-range Lightning F Mk 6s. All are capable of inflight refueling. France's air defense command can operate within the NATO air defense ground environment, although France has no forces assigned to the alliance in peacetime. The Armée de l'Air's interceptor force comprises four wings operating the Mirage IIIC and Mirage F1 interceptors. These are to be joined by the Mirage 2000 later in the 1980s. Spain's air defense command operates a single wing each of the Mirage III, Mirage F1 and F-4C Phantom.

If the NATO powers have in general neglected air defense forces, the Soviet Union by contrast has devoted considerable resources to this task. The interceptors, surface-to-air missiles and ground environment of the Soviet air defenses constitute an independent service, quite distinct from the air force. It is the largest air defense force in the world, with 10,000 SAMs, over 5000 early-warning and height-finding radars and 2500 manned interceptors. Although considered effective against aircraft operating at medium and high altitudes, it is less well-equipped to counter low-flying aircraft and cruise missiles. However, systems now coming into service are intended to make good this deficiency. In addition to its air defenses, the Soviet Union has also deployed a ballistic missile defense system around Moscow with 32 Galosh ABMs.

The air defense forces are organized into ten districts, each covering a specific area of the Soviet Union. Defenses are concentrated around industrial centers, strategic forces' bases and other military targets. The bulk of the interceptor regiments are deployed to face the threats from NATO and the People's Republic of China. The air forces of the Warsaw Pact allies defend a buffer zone between the Soviet Union and the European NATO powers and are mainly equipped with interceptors for this purpose. The most numerous interceptor aircraft with the Soviet air defense regiments is the MiG-23 Flogger. They also fly Sukhoi Su-9 and Su-11 Fishpots, the Su-15 Flagon single-seat interceptors; the high altitude MiG-25 Foxbat; and long-range Tu-28P Fiddler and Yak-28 Firebar.

Mainstay of the USAF's interceptor force, the Convair F-106A Delta Dart first entered service in 1959. It is a single seat, all-weather interceptor of tail-less delta configuration. Power is provided by one 24,500lbs afterburning Pratt & Whitney J75-P-17 turbojet, which gives the F-106A a maximum speed of Mach 2.3 at altitude. The F-106A spans 38ft 3in and is 70ft 7in long, with a takeoff weight of 38,700lbs. Rate of climb from sea level is 39,800ft per minute and combat ceiling is 52,000ft.

The F-106A is fitted with a highly-automated MA-1 weapons control system, which directs interceptions and weapons release. Armament comprises a single AIR-2A Genie unguided air-to-air rocket, which carries a nuclear warhead and has a range of about six miles, plus four AIM-4 Falcon AAMs. These can either be AIM-4Fs with semi-active radar guidance, or AIM-4Gs with IR homing, both having a range of seven miles. The missiles are carried in an internal weapons bay and, as first produced, this was the F-106's sole armament. However, as with the F-4 Phantom an internal gun armament was later thought to be necessary and a 20mm M-61 cannon was fitted under the Sixshooter modernization program.

The F-106's stablemate within ADCOM for many years, the McDonnell F-101B and F-101F Voodoo is now on the point of retirement. It is a two-seat, interceptor powered by 14,990lbs afterburning Pratt & Whitney J57-P-55 turbojets giving a maximum speed of Mach 1.85. Rate of climb from sea level is 36,000ft per minute and combat ceiling is 51,000ft. Armament comprises two AIR-2A Genie rockets and two AIM-4C IR guided Falcons. The Canadian Armed Forces' CF-101B is essentially similar.

France's Mirage F1C all-weather, single-seat interceptor differs from other members of the Mirage family in abandoning the tail-less delta formula for a high-mounted swept wing and horizontal tail surfaces. It is powered by a SNECMA Atar 09K-50 turbojet, which develops 15,875lbs thrust with afterburning. The Mirage F1 spans 27ft 7in, is 49ft 2in long and has a maximum takeoff weight of 32,850lbs. Performance includes a maximum speed of Mach 2.2 and a service ceiling of 65,000ft. Built-in armament consists of two 30mm cannon. Up to three Matra R530 or Super R530 AAMs, with alternative IR or semi-active radar guidance, can be carried, with Sidewinder or Matra Magic AAMs mounted on the wingtips. The Mirage F1's replacement will be the Mirage 2000 delta-winged interceptor, which will have a lookdown/shootdown capability.

The United Kingdom's primary interceptor is the ubiquitous Phantom, but two squadrons of the older BAe Lightning F Mk 6 remain in service. This swept-wing, single-seat fighter is powered by two Rolls-Royce Avon Series 300 turbojets, each delivering 16,300lbs thrust with afterburning. Unusually the engines are mounted one above the other, rather than side-by-side. The Lightning spans 34ft 10in, is 55ft 3in long and has a loaded weight of 49,000lbs. Maximum speed is over Mach 2 and ceiling is 60,000ft. A weakness of the design is its limited endurance, which has been somewhat alleviated in the F Mk 6 by fitting a large ventral fuel tank. Armament comprises two 30mm Aden cannon and two IR homing Red Top AAMs of seven miles range. In an emergency the RAF plans to use its Hawk T Mk 1 weapons trainers as point defense interceptors, armed with Sidewinder AAMs.

The RAF's next interceptor, due to enter service in 1985, is the Panavia Tornado F Mk 2, the air defense variant of the Tornado GR Mk 1. This two-seat aircraft is generally similar to the Tornado GR 1, but it has an elongated nose housing a Ferranti Foxhunter air interception radar with a range of some 100 nautical miles. Performance includes a maximum speed of Mach 2.2 and a maximum ferry range (with full fuel load) of over 2000 miles. Built-in armament consists of a single 27mm Mauser cannon, with two Sidewinder short-range AAMs on wing pylons. Four medium-range Skyflash AAMs, which have the ability to snap up or down to engage both low and high flying targets, are recessed under the fuselage.

Interestingly the variety of interceptors produced by the Soviet Union alone is almost as great as those built by the various NATO nations. The MiG-23 interceptor is the same aircraft as that flown by Frontal Aviation. It is now, in terms of numbers in service, the most important aircraft in the Soviet air defense force. In the 1960s this position was held by the now-superseded MiG-21, which, however, still serves as an interceptor with the Warsaw Pact satellites, and as a tactical fighter with Frontal Aviation.

The delta-wing Su-15 Flagon, unlike the MiG-21 and MiG-23, is a specialized interceptor unsuited to any other role. Some 700 of these single-seat, twin-engined fighters serve with the Soviet air defenses. The Flagon was originally envisaged as a progressive development of the earlier Su-9 and Su-11 Fishpot, with a larger and more capable radar. However it emerged as a considerably heavier aircraft, powered by two Tumansky R-13 turbofans developing 16,000lbs thrust with afterburning. Maximum speed is over Mach 2 at altitude, service ceiling is 55,000ft and combat radius 400 miles. Maximum takeoff weight is around 45,000lbs and dimensions include a span of 34ft 6in and a length of 70ft 6in. Armament consists of two 14 mile range AA-3 Anab AAMs carried underwing, one infra-red guided the other semi-active radar homing. Soviet tactics require interception sorties to be flown under strict ground control and this rigid doctrine is reflected in the design of the Su-15, which has little dogfighting maneuverability in contrast to the agile F-106.

The Su-9/Su-11 Fishpot, precursor of the Flagon, is still in service, with perhaps as many as 600 being operational. Powered by a 22,000lbs afterburning Lyulka AL-7F turbojet, the Su-9 has a maximum speed of Mach 2.1 and a service ceiling of 65,000ft. Its delta wing spans 27ft 8in and overall length is about 60ft. The Su-11 is essentially similar, except that a modified nose houses a more powerful Skip Spin radar. Because of its less capable radar the Su-9's armament is limited to four AA-1 Alkali 'beam-riding' AAMs, whereas the Su-11 carries two semi-active homing Anabs.

In many ways the most impressive interceptor in current service with the Soviet air defenses, the MiG-25 Foxbat-A was almost certainly designed to combat the

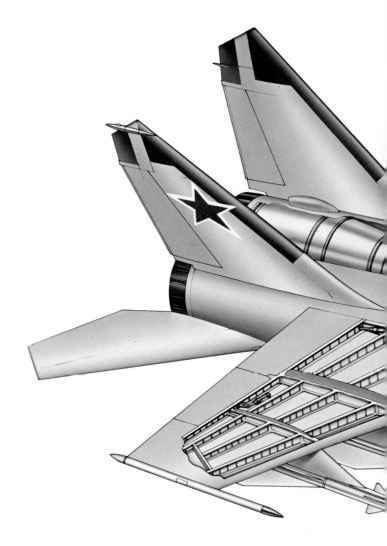

fast, high-flying B-70 Valkyrie bomber. The MiG-25 is powered by two Tumansky R-31 turbojets, which give 27,000lbs of thrust each with afterburning. Maximum speed in clean condition is Mach 3, which is reduced to Mach 2.8 when external stores are carried. Ceiling is in the region of 75,000ft and the initial rate of climb is 30,000ft per minute. Maximum takeoff weight is 82,500lbs, the span is 46ft and length 73ft 2in. Operating at up to 250 miles from base, the MiG-25 can carry four AA-6 Acrid AAMs, a mix of two IR and two semi-active radar homing being typical. The semi-active homing AA-6 has a range of some 50 miles, while the IR version's range is half this.

Notwithstanding its impressive performance, the Foxbat is to a large extent an aircraft without a mission. SAC never received the B-70 and instead adopted low-level penetration tactics. The counter to these is an interceptor with radar and missiles having a lookdown/shootdown capability. These characteristics have been built into a new two-seat variant of the MiG-25, code-

named Foxhound, which entered service in 1982. It has a radar reputedly based on F-14 Tomcat technology acquired from Iran and snap-up/snap-down AA-9 AAMs.

Long range is always desirable in an interceptor, but for the Soviet Union with its long Arctic frontier it is of paramount importance. So it is not surprising that the largest fighter in service anywhere – the Tupolev Tu-28P Fiddler – guards these inhospitable wastelands. With a maximum takeoff weight of 85,000lbs, the Tu-28P spans some 65ft and is 85ft in length. Two crew members are carried, seated in tandem in the forward fuselage behind the powerful 'Big Nose' radar. The powerplant is two Lyulka AL-21F turbojets, developing 24,500lbs of thrust with afterburning. Maximum speed is Mach 1.8 and service ceiling 60,000ft. Tactical radius is some 800 miles. The all-missile armament comprises four AA-5 Ash missiles of 18 miles range carried underwing. Their guidance can either be IR or semi-active radar homing, but the latter system is probably of shorter range.

The Soviet MiG-25 interceptor (cutaway artwork) is capable of speeds of Mach 3. The Hawk mobile SAM (below) is widely used by the US forces and the NATO allies. The British Rapier (top) is especially valuable for airfield defense. The Alpha Jet (above) serves with Groupement Ecole 314 at Tours, the Armée de l'Air's fighter pilot training school.

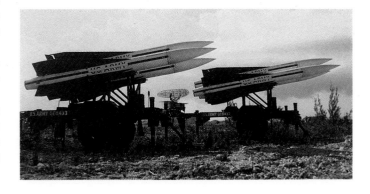

The Yak-28P Firebar fulfils the same role as Fiddler in the less important Soviet air defense districts. In performance it falls between the Tu-28P and Su-15. A swept-wing aircraft, the Firebar has two 13,000lbs Tumansky R-11 afterburning turbojets, one mounted under each wing. Maximum speed is Mach 1.3 and service ceiling is 55,000ft. It is armed with two AA-3 Anab AAMs, in the usual mix of one IR and the other radar guided. Soviet tactics may require that those missiles are salvoed simultaneously to present the target bomber with a difficult problem in countermeasures. A new Soviet long-range interceptor is due to enter service by 1983. A twin-engined aircraft with two crew members, it has provisionally been identified as the Su-27.

Air-to-air missiles for the air defense mission are in general larger and longer ranging than those carried by the air superiority fighter. However many missiles have been found to be useful in both roles, notably the AIM-7 Sparrow. The most sophisticated AAM in service anywhere in the world is undoubtedly the AIM-54 Phoenix, carried by US Navy's F-14 Tomcat. Its range is over 100 miles, with most of the flight controlled by semi-active radar homing and an active radar seeker providing guidance for the last 12 miles. The AIM-54A's speed is over Mach 5 and the warhead is 130lb high explosive with proximity and impact fuzing. The Soviet AA-6 Acrid carried by the MiG-25 is a Mach 4.5 missile with a range of up to 30 miles. However, this limitation is due to the tracking range of the Foxfire radar and missile ranges of up to 50 miles may be possible with an improved radar.

A serious problem arising from the use of long range missiles is the positive identification of the target as hostile. The standard IFF equipment, used by air-to-air and ground-to-air weapons systems, interrogates a radar contact electronically. If the aircraft is friendly a coded response is returned. However, the lack of a correct response cannot alone be positive proof that the radar contact is hostile. It may be a neutral, or a friendly aircraft with unserviceable IFF equipment.

One way to resolve this is for interceptors to work in pairs. The leader flies past the target at high speed and radioes its identity to his wingman. If it is hostile, the leader breaks away sharply, allowing the wingman to attack. This tactic is unsatisfactory for several reasons. It is wasteful of scarce and expensive interceptor aircraft, requiring two to do the job of one. It can also alert an otherwise unsuspecting enemy and put friendly aircraft at risk.

The long-term solution seems to be to fit the interceptor with an electro-optical magnification device 'slaved' to the radar, to allow visual identification at increased ranges. An interim arrangement, which has

The Soviet Union deploys an extensive range of ground-based air defense systems, including the Galosh anti-ballistic missile (top left) and the SA-3 Goa (left), a mobile medium-to-low altitude surface-to-air missile.

been successfully tried on F-14s and F-15s, is to mount a telescope alongside the HUD. This allows an F-14-sized target to be picked up at 22 miles and positively identified at about half this distance. The more sophisticated magnification systems include the USAF's TISEO (target identification system, electro-optical), which is already fitted to some F-4Es, and the Tornado F Mk 2's VAS (visual augmentation system). The range of these can be 30 miles in ideal conditions.

Interceptors' radars should have good range, the ability to deal with several targets at the same time, resistance to ECM jamming and be capable of picking out low-flying targets as small as cruise missiles from the clutter of ground radar returns. It is symptomatic of the neglect of air defense forces by the USAF that the most capable system currently in US service is the AWG-9 fitted to the Navy F-14A Tomcat. It has a maximum range of 170 nautical miles against a bomber size target and can track up to 24 targets at once and simultaneously engage six of them. The Foxhunter radar fitted to the British Tornado F Mk 2 is less costly than the AWG-9. It has a 100 nautical mile range and can handle multiple targets. Resistance to ECM jamming is especially good.

The MiG-25's Foxfire radar's 54 nm search range does not compare very favorably with the latest Western types. However, as Soviet interceptors operate under rigid ground control, maximum radar range is not so important as for NATO interceptors. In general Soviet radar technology lags behind that of the NATO powers, the Foxfire lacking such refinements as transistors and printed circuitry. However, resistance to jamming is considered to be good. The MiG-23's High Lark radar has a search range of 46 nautical miles and that for the Skip Spin fitted to the Su-15 and Yak-28 is 22 nautical miles. The Tu-28P's Big Nose radar's search range is 32 miles.

While the interceptor is the main offensive weapon of air defense forces, providing defense in depth, the surface-to-air missile is useful as a last ditch defense and for point protection of such high value targets as airfields. Mobile SAMs also contribute to the air defense of armies in the field. SAMs normally operate in a predesignated missile zone, backing up an air interception zone.

Missile belts are part of the air defenses on NATO's Central Front in Europe, the standard area defense missiles being the MIM-14B Nike Hercules and MIM-23B Improved Hawk. Nike Hercules is a long-range SAM, which for many years was deployed as part of the North American air defenses. It was phased out of American service in the 1970s but remains in use by the United States' European and Far Eastern allies. The MIM-14B is a two-stage missile powered by solid propellant. Its length is 41ft and launch weight is 10,000lbs. Either a conventional or nuclear warhead can be fitted and range is about 90 miles.

The Hawk low and medium altitude SAM is widely

used by the US armed forces and the NATO allies. Most Hawks in current service are the MIM-23B Improved Hawk, which has a maximum range of 25 miles. The major modifications incorporated into the Improved Hawk are a new semi-active radar guidance system, a larger warhead and improved propellant. The countries operating this system include Germany, France, Italy, Denmark, Belgium, the Netherlands, Greece and Spain.

The UK has deployed the Bloodhound Mk 2 50-mile range SAM as an area defense weapon to guard the vulnerable East Anglian air bases. This is backed up by the shorter range BAe Rapier which provides low-level, point defense for the RAF and USAF airfields in the UK and also the bases in RAF Germany. Rapier's range

is about four miles and guidance is command to line of sight, the operator keeping his sight on the target onto which the missile is then automatically guided.

The Soviet Union makes extensive use of SAMs for air defense, quite apart from the mobile tactical systems which provide air cover for armies. The elderly SA-1 Guild provides the primary missile defense for Moscow, with two rings of launchers deployed around the city. The SA-1 first entered service in the mid-1950s and is believed to use active radar homing. Range has been assessed at about 30 miles.

The SA-2 Guideline has for long been the standard Soviet air defense SAM. It is deployed throughout the USSR and has been exported. It was used extensively by the North Vietnamese during the Southeast Asia

The US air defenses are to be improved with E-3A AWACS aircraft (top), supplementing the ground radars of the DEW line stations (above).

conflict. It is a medium altitude missile, with a range of 30 miles. As more SA-10 missiles are deployed, the SA-2 force will be gradually reduced. The SA-10 is the latest strategic air defense SAM to be deployed by the Soviet Union. It has a range of over 60 miles and is effective at low, medium and high altitudes. In view of the current SAC penetration tactics, the low altitude capability of the SA-10 is especially valuable and it may also be effective against cruise missiles.

Medium to high altitude coverage is provided by the SA-5. This 185 mile range missile entered service in 1963 and is now operational at more than 100 launch complexes with deployment continuing. It is complemented by the SA-3, a low to medium altitude point defense missile of 12 mile range. This SAM is deployed to cover targets of special importance and there are some 400 launching sites. A program is now in progress to convert the SA-3 launchers from two- to four-rail mountings.

The most exotic surface-to-air missiles are undoubtedly the anti-ballistic missile systems. The United States' Safeguard site was dismantled in 1975. This used the 460-mile range Spartan missile for destruction of incoming missiles or warheads outside the earth's atmosphere. It was backed up by the shorter-range (25 miles) Sprint, which was a fast reaction missile intended to pick off those warheads which had evaded the Spartans. Both missiles carried nuclear warheads. Research into ABM systems continues and they may be reintroduced to defend MX launch sites. The Soviet Union at present deploys 32 AMB-1b Galosh missiles in defense of Moscow (up to 100 are allowed under the 1972 SALT treaty). Two new ABMs are under development, SH-4 and SH-8, similar in concept to the Spartan/Sprint combination and the Soviet Union continues to develop and deploy the sophisticated radars on which the ABMs depend for targeting information.

The interceptors and SAM forces are the offensive elements of air defense, but they cannot function effectively without the information provided by ground based radars and control centers. The air defense ground environment (ADGE) in its turn is dependent on good communications to relay data to interceptors and missile sites and also to exchange information between control centers. The ADGE has been extended into space to provide early warning of ballistic missile attack and surveillance of the numerous space satellites.

The North American continent's first line of defense against attack by aircraft or cruise missiles is the DEW (distant early-warning) Line of radars which stretches from Alaska to Greenland. There are at present 31 radars in this chain, but the number is being reduced to 13. All are linked to the NORAD Combat Operations Center at Cheyenne Mountain. Improvements will replace the existing radars with a mixture of long-range radars, with shorter-range unattended radars filling gaps in the coverage. A second radar

chain, the Pine Tree Line, follows the border between the United States and Canada.

Further coverage of North American airspace is provided by the Joint Surveillance System, which is shared by the Federal Aviation Authority and USAF. In all there are 83 sites in this network. The seaward approaches to the United States are to be covered by two over-the-horizon backscatter (OTH B) radars from the mid-1980s. This advanced system, which has yet to be proved operationally viable, will provide all-altitude detection out to 1000 nautical miles from the coast. It achieves this outstanding range by bouncing radar waves off the ionosphere, a technique which is not usable in polar regions to replace the DEW Line radars. Cuba presents a particular problem to the US air defenses, as a Soviet ally within the Western Hemisphere. Consequently a balloon-borne long-range radar, based at Cudjoe Key, Florida, maintains a special watch in this direction.

Early warning of missile attack is provided by three BMEWS (ballistic missile early warning system) radars at Clear, Alaska, Thule, Greenland, and Fylingdales in the United Kingdom. Early warning satellites are also used for this purpose and the Safeguard system's perimeter acquisition radar remains operational and is capable of detecting incoming ICBM warheads. Submarine-launched ballistic missiles can be detected by two Pave Paws radars covering the Atlantic and Pacific. A further shorter range FSS-7 radar covers the Gulf of Mexico. As well as maintaining a ballistic missile early warning system, NORAD also keeps track of all objects in space following an earth orbit. Information is derived from radar and optical tracking sites in the USA and various overseas locations scattered from New Zealand to Turkey.

The European NATO allies maintain a network of surveillance radars and associated command and communications systems. The coverage of the NATO air

The Soviet AEW aircraft is the Tu-126 Moss (above left). The RAF's Nimrod AEW Mk 3 (left and above) is due to enter service in 1983.

defense ground environment (NADGE) embraces Norway, Denmark, West Germany, the Netherlands, Belgium, Italy, Greece and Turkey. The separate UKADGE is linked to this network and France also exchanges information with the other NATO members, but does not allow them to control her interceptors or missiles. The NATO systems are designed to cope with high levels of air activity and they use automatic data processing to improve reaction times. All control centers are connected and interoperable so that information is shared. NATO is also aware of the threat to its radars from enemy action and some of these are mobile units, which are more difficult to target.

The greatest weakness of a ground based radar is its limited detection range against low flying aircraft. Radar beams follow a straight line and, as the earth's surface is curved, a low-flying aircraft soon falls below the horizon. Even under optimum conditions, detection range will only be some 60 nautical miles, giving only a short time to react to a threat. Hence the development of AWACS (airborne warning and control system) aircraft, which, by carrying the radar aloft, greatly extend the coverage against low-flying aircraft. An AWACS flying at 30,000ft can detect a low-flying aircraft at more than 200 nautical miles range. Furthermore it can, in theory at least, pick up an aircraft flying at its own altitude at double this distance.

The USAF fulfils this mission with the Boeing E-3A Sentry, a derivative of the Boeing 707 airliner which carries its APY-1 radar in a large rotating radome atop the fuselage. The USAF plans to buy at least 40 E-3As

and NATO will acquire a further 18. With a span of 145ft 9in and length of 152ft 11in, the E-3A has a maximum takeoff weight of 325,000lbs. Power is provided by four Pratt & Whitney TF33-PW-100 turbofans of 21,000lbs thrust each. Maximum speed is 530mph and service ceiling 35,000ft. Endurance is over 11 hours and this can be extended by inflight refueling. The normal crew complement is 17. The immense amount of information gained from the radar is handled by operators using computer processing. Data can be transmitted to ground stations or interceptors by voice radio and data link. The E-3A is fitted with the Joint Tactical Information Distribution System (JTIDS), which can provide secure communications between all air defense elements.

The UK has its own early-warning aircraft, the Nimrod AEW Mk 3, which is due to become operational in 1983. It is a conversion of the maritime reconnaissance Nimrod MR Mk 1, fitted with radar antennae in the nose and tail. Eleven aircraft will replace the elderly Shackleton AEW Mk 2s which currently provide the RAF with AEW. The Nimrod AEW Mk 3 will be fully interoperable with the NATO E-3A fleet and compatible with the NATO air defense ground environment. It differs from its American counterpart in being designed from the outset to carry out maritime surveillance, as well as AEW roles. However, the E-3A has been modified to give it a measure of sea surveillance capability.

Although it is often regarded as the Soviet equivalent to the E-3A, the Tupolev Tu-126 Moss is in fact a much earlier aircraft. It became operational in about 1970, some two years before the RAF's Shackleton AEW Mk 2s, and the first E-3As were not delivered until 1977. Therefore it is unlikely that the Soviet aircraft would be as capable as the E-3A, even discounting the acknowledged Soviet backwardness in electronics. The Tu-126's ability to detect low flying aircraft is assessed as only marginally effective over water and nonexistent over land. However, this may be an evaluation of the aircraft in a role for which it was never designed. Perhaps the Moss is simply an airborne radar picket, intended to extend the range of ground based radars against aircraft at medium and high altitudes. As such it would complement the Tu-28P in operations over the Soviet Union's Arctic frontiers.

The Tu-126's design is based on that of the Tu-114 airliner, with its radar mounted above the fuselage in an installation similar to that of the E-3A. Power is provided by four Kuznetsov NK-12MV turboprops of 15,000shp, giving a maximum speed of 460mph. Span is 167ft 8in and length 188ft, with a takeoff weight of 365,000lbs. Service ceiling is 33,000ft and endurance is an incredible 20 hours, which can be further extended by inflight refueling. Only a small number of Tu-126s are in service (less than 20). A new Soviet AEW aircraft is reported to be under development, possibly based on the Ilyushin Il-86 wide-bodied airliner.

An RAF Jaguar GR Mk 1 fitted with a reconnaissance pod flies over the West German countryside. This aircraft serves with No 2 Squadron at Laarbruch.

6. RECCE AND ELINT

Recce and ELINT

Air reconnaissance is an indispensible preliminary to many military operations, as the knowledge it makes available about enemy forces and intentions may form the basis of a commander's plan of action. Tactical air reconnaissance is primarily intended to provide timely information of immediate use to the commander on the battlefield. However, the longer term aims, of building up an overall view of an enemy's military strength and industrial capacity so that his war-fighting capability may be accurately assessed, are equally important. This strategic reconnaissance task may also be carried out by aircraft, although air reconnaissance is only one of many information sources available to intelligence officers.

Reconnaissance aircraft use many different sensors to carry out their diverse missions. The visual observation of the crew is often valuable and their comments can be recorded for later analysis. The camera is of course still a primary reconnaissance sensor and it is used both for area and pin-point coverage. Imagery can be recorded on conventional photographic film, or an infra-red picture can be obtained which will reveal much that is hidden from the naked eye. A further alternative is to record a radar picture. This data can be stored aboard the reconnaissance aircraft until its return to base, or transmitted direct to a battlefield headquarters, thus giving a commander 'real time' information. Reconnaissance is not confined to visual imagery, however, as electronic intelligence (ELINT) is a very important aspect of this activity. ELINT aircraft monitor enemy radio and radar transmissions, so that the performance of equipment and operational procedures can be assessed and an electronic order of battle can be built up.

The introduction of reconnaissance satellites has tended to diminish the importance of the strategic reconnaissance aircraft, but these nevertheless continue in service. This is because they can be speedily deployed to cover unexpected contingencies and can reconnoiter areas which would not justify the expense of satellite coverage. Since Powers' U-2 was shot down in 1960, it is unlikely that American aircraft have overflown Soviet territory, but they certainly operate around the Soviet borders using stand-off systems to gather intelligence. They have also covered the territory of other potential adversaries, an activity publicized in 1981 when North Korean SAMs unsuccessfully attempted to shoot down a Lockheed SR-71.

The United States' strategic reconnaissance aircraft operate within Strategic Air Command, as one of their functions is to provide the command with targeting information. The 55th Strategic Reconnaissance Wing flies Boeing RC-135 aircraft, primarily used for ELINT, from Offutt AFB, Nebraska. The 9th SRW is based at Beale AFB, California, and operates one squadron of Lockheed SR-71A Blackbirds and a second of Lockheed U-2s. The 9th SRW has detach-

ments flying from Mildenhall in the UK and Kadena AB on Okinawa. Other airfields are regularly used for strategic reconnaissance flights, including Hellenikon AB in Greece and Eielson AFB in Alaska.

The Lockheed SR-71 Blackbird is a high-performance, high altitude reconnaissance aircraft, which carries a crew of two, the pilot and reconnaissance systems officer (RSO). It has a loaded weight of some 170,000lbs and spans 55ft 7in, with a length of 107ft 5in. Power is provided by two Pratt & Whitney JT11D-20B turbojets, which can operate continuously on afterburner producing 32,500lbs of thrust each. Maximum speed is over Mach 3 and operating altitude is between 80,000ft and 100,000ft. Maximum fuel load is some 80,000lbs of special JP-7 and the Blackbird operates with KC-135Q tanker aircraft. The nose-mounted sensors can cover 100,000 square miles in one hour and can be changed to suit the mission. A typical sortie will last for 2½ hours, but this can be extended to four hours giving a mission radius of some 4000 miles.

The SR-71's high operating altitude and Mach 3 plus speed make great demands on crew and aircraft alike. The pilot and RSO wear astronaut-style pressure suits for safety in the aircraft and in the event of an ejection. The SR-71 requires great skill from its pilot since very precise flying at Mach 3 is virtually impossible – at this speed the radius of a turn is 90 miles. High airspeeds also cause the aircraft's skin to heat, with temperatures varying from 450 degrees Farenheit up to 1100 degrees F. The aircraft's distinctive black paint scheme helps deal with this problem, as it radiates heat more effectively than natural metal, and the fuel is also used for cooling. After landing the aircraft is literally

The Lockheed U-2 strategic reconnaissance aircraft has been in US service since the late 1950s, an early U-2B being pictured (above left). The USAF's standard tactical reconnaissance aircraft is the RF-4C Phantom, an aircraft of the 10th TRW based at Alconbury in the UK is pictured (above).

too hot to touch and because of the heating problem over 90 percent of the structure and skinning is titanium. The engines too are quite distinctive, as they operate as turbojets at low airspeed and as ramjets at high Mach numbers.

The SR-71 originated as the YF-12A experimental interceptor, which first flew in 1962. The project was highly classified and the SR-71's existence was not officially confirmed until 1964. Blackbirds entered service in January 1966 and it is believed that 31 were built. Two of these were converted to SR-71B pilot trainers and one of these has been lost in an accident.

The Lockheed U-2, which entered service with the USAF in 1957, still equips the 9th SRW's 99th Strategic Reconnaissance Squadron. A number of the squadron's aircraft are U-2Cs, which are used for electronic reconnaissance. This version is powered by a 17,000lbs thrust Pratt & Whitney J75-P-13 turbojet, which gives a cruising speed of 460mph. Operating altitude is about 80,000ft and range (which is well over 4000 miles) can be increased by gliding flight between intermittent bursts of engine power. The high aspect-ratio, glider-like wing spans 80ft and the fuselage length is 49ft 7in. Electronic reconnaissance equipment is carried in an 8ft long canoe-like fairing atop the fuselage. Most of the current U-2 fleet are U-2R variants, a much larger aircraft than the C, with a

lengthened fuselage and larger wing, which is increased in area by over 70 percent. The nose section houses various interchangeable reconnaissance packs. Twenty-five U-2Rs have been produced by rebuilding earlier airframes.

The MiG-25 represents the Soviet Union's equivalent to the SR-71 and two reconnaissance versions, Foxbat-B and Foxbat-D are currently in service. The aircraft can operate as high as 80,000ft at a maximum speed of Mach 3.2. Shortly after its introduction into service in 1971, four Foxbat-Bs were detached to Egypt and flew reconnaissance missions over the Israeli-occupied Sinai peninsula without being intercepted. Similar detachments have operated from Syria and Libya and Soviet-based Foxbat-Bs have overflown Turkey and Iran. The aircraft has also made shallow penetrations of NATO airspace, overflying West Germany, Denmark and Norway. However with the deployment to Europe of the F-15, the MiG-25 has lost much of its immunity, but it continues to operate effectively over less-well defended areas and as a stand-off system.

Foxbat-B differs little from the interceptor MiG-25 apart from equipment changes for its new role. The large Foxfire radar is replaced by a battery of cameras, a sideways-looking airborne radar (SLAR), plus ground-mapping doppler and forward-looking radars for navigation. The cameras are installed in oblique pairs, angled to port and starboard at angles of 15 and 45 degrees, with the fifth mounted vertically. Foxbat-D is equipped primarily for ELINT and has a larger SLAR, in addition to electronic monitoring equipment. Some 150 Foxbat-B and D reconnaissance aircraft have been

produced for service with Frontal Aviation.

Until the early 1970s the Soviet Union also operated the Yak-25RD Mandrake, a high-flying, single seat strategic reconnaissance aircraft. This had a 72ft span unswept wing and operated above 60,000ft over a range of more than 2000 miles. Recent reports have indicated that a belated successor to Mandrake is being tested. Identified as Ram-M, this aircraft is broadly similar to the Lockheed U-2, but has twin vertical tail surfaces. Echoing the practice of both Britain and Germany before World War II, the Soviet Union apparently uses civil flights by aircraft of Aeroflot for clandestine reconnaissance missions. In November 1981 a Soviet airliner overflew the Groton Shipbuilding Yard in Connecticut, where Trident submarines are made.

In contrast to the highly-specialized strategic re-

connaissance aircraft, tactical reconnaissance machines are generally modifications of standard fighter and attack aircraft. This allows these aircraft to undertake a secondary strike/attack role and simplifies maintenance, by keeping the number of aircraft types in service to a minimum. Some air forces take this philosophy a stage further and simply install self-contained reconnaissance pods on otherwise unmodified aircraft. Although this simplifies switching the reconnaissance aircraft to another role it does not solve the problem of crew training. Tactical reconnaissance is a highly skilled operation and aircrew proficient in the art will have little time to train for an equally specialized secondary mission. The best compromise is to assign tactical reconnaissance units a secondary tactical nuclear strike role, as this mission most closely resem-

The SR-71 Blackbird (overleaf) is a Mach 3 plus strategic reconnaissance aircraft, which serves with the USAF's 9th SRW at Beale AFB, Ca. One of most useful reconnaissance sensors is infra-red linescan (right), which can produce a thermal picture of the terrain overflown by day or night (below). The most important Soviet tactical reconnaissance aircraft is the Fishbed-H version of the MiG-21, shown (below right) in Polish service.

bles their primary task.

The USAF's standard tactical reconnaissance aircraft is the RF-4C version of the Phantom. A total of 505 was built between 1964 and 1974 and the aircraft now serves with TAC's 363rd Tactical Reconnaissance Wing at Shaw AFB, SC, and the 67th TRW at Bergstrom AFB, Texas. The 363rd TRW is scheduled to become a tactical fighter wing with the F-16, but it will retain a single squadron of RF-4Cs. In Europe the RF-4C serves with the 10th TRW at Alconbury in the UK and 26th TRW at Zweibrücken in West Germany. One squadron of the 18th TFW at Kadena, Okinawa, operates the RF-4C in the USAF's Pacific Air Forces.

The first reconnaissance version of the Phantom was the US Marines' RF-4B, a straightforward conversion of the F-4B with a modified nose housing vertical and oblique cameras. The RF-4C, based on the F-4C, has a more comprehensive range of sensors. There are three camera stations, that in the nose usually being a forward-looking oblique camera, with a fan of three low-level cameras in the center position and a panoramic camera mounted vertically aft. However, there are numerous permutations, which can be matched to high or low level missions by day or night. For night photography photoflash cartridges are ejected from the rear fuselage to provide illumination.

The RF-4C also carries an infra-red linescan sensor, which records the heat signatures of objects on the ground. It can, for example, penetrate camouflage and locate hidden vehicles. IR can show which aircraft on an airfield have recently been fueled or had their engines running. It can even reveal traces of aircraft which have recently taken off, as their presence remains as a heat shadow on the hardstanding. Radar reconnaissance is undertaken by a nose-mounted forward-looking radar and a sideways-looking airborne radar, which scans to either side. The SLAR is able to pick out moving targets, such as truck convoys, using doppler techniques. The RF-4C is currently being modified to incorporate a data link, to transmit the SLAR picture to a ground station 30-50 nautical miles away. This eliminates the delays in getting the results of a sortie to the commanders who requested it.

Accurate navigation is essential for effective tactical reconnaissance and the RF-4C is accordingly fitted with an inertial navigation set, navigation computer and radar altimeter. Some aircraft also have the very accurate LORAN long-range navigation equipment. A data annotation device records the aircraft's position, altitude and other performance parameters on the film as it is exposed. The RF-4C carries no defensive armament, but can be fitted with ECM jamming pods. It can also carry a tactical nuclear weapon on the centerline station for its secondary strike role. Otherwise the external stores stations carry auxiliary fuel tanks with a total capacity of 1340 gallons, supplementing the internal tankage of 1889 gallons. This gives a radius of 513 miles for a low-level mission, or 673 miles at high altitude.

The USAF is to supplement the RF-4C force with the Lockheed TR-1A, a stand-off, tactical reconnaissance aircraft based on the U-2. A total of 35 is on order and one squadron will deploy to Alconbury in the UK to cover the NATO central region. The TR-1A spans 103ft and is 63ft in length. Power is provided by one Pratt & Whitney J75-P-13B, which gives a cruising speed of 430mph. The aircraft can operate at altitudes of 90,000ft and range is over 3000 miles. Operating at high altitudes, the TR-1A's advanced, synthetic aperture, sideways-looking radar can detect ground activities several hundred miles to the side of the aircraft. Therefore a TR-1A operating over NATO territory can warn of movements deep in Warsaw Pact territory, without having to evade the enemy defenses. The aircraft carries a variety of interchangeable sensors in the nose bay and in wing pods, the maximum payload being almost two tons. The aircraft's endurance is up to 12 hours, although sorties of this length would make great

Reconnaissance versions of the ubiquitous Canberra (below) remain in service with several Third World air arms and the RAF maintains a flight of Canberra PR Mk 9s. The Mirage IIIR (above) serves with the Armée de l'Air's 33ᵉ Escadre de Reconnaissance at Strasbourg. The Swedish air force maintains a number of obsolete Lansen fighters (right) in service for target towing and ECM duties.

A Canadair CL-289 reconnaissance drone (top) is launched from its ground vehicle. After following a preprogrammed flight path to its objective, it is recovered by parachute (right), automatically-inflating airbags absorbing the landing impact. The Canadair CL-227 (above) is a rotary-wing reconnaissance drone, which can carry a TV camera, infra-red sensor, or laser target designator, on missions lasting from two to three hours.

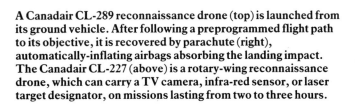

demands on the pilot, who has to wear an uncomfortable pressure suit throughout the flight.

Many of the European NATO Allies make use of pod-mounted reconnaissance systems flown on tactical fighters. A typical example is the RAF's tactical reconnaissance Jaguar, which carries a pod-mounted camera and IR-linescan. Two RAF Jaguar squadrons specialize in reconnaissance; No 2 Squadron at Laarbruch in RAF Germany and No 41 at Coltishall in the UK. The aircraft are standard Jaguars, the inertial navigation system being especially useful. Reconnaissance cameras are housed in two rotatable drums within the pod, which swivel to expose the camera ports during photography. The front drum contains two side-mounted and one forward-looking oblique camera, while the second drum has two oblique cameras for low-level work, or a single vertical camera for use at medium altitudes. This combination gives horizon-to-horizon coverage and a data conversion unit obtains the aircraft's position from the navigation computer and annotates it onto the film. The IR-linescan is especially useful in poor light and darkness and it too has the aircraft's position automatically marked on the film. At the end of a sortie the film is rapidly processed so that within ten minutes it is being examined by interpreters. The results of their examination, together with the pilot's report, can be passed on within half an hour of the aircraft landing.

France's Armée de l'Air has a single reconnaissance wing, the 33e Escadre de Reconnaissance based at Strasbourg, with three component squadrons flying Mirage IIIR and Mirage IIIRD aircraft. These are variants of the Mirage IIIE multi-role fighter, fitted with five nose-mounted cameras suitable for high, medium and low altitude reconnaissance. The Mirage IIIRD is more advanced, with IR sensors and doppler-navigation equipment. From 1983 both types will be replaced by the Mirage F1CR (62 on order). This reconnaissance version of the F1C interceptor has provision for inflight refueling and carries cameras, IR sensors, forward-looking radar and an inertial navigation system.

West Germany's Luftwaffe operates the RF-4E version of the Phantom, 88 of which were delivered in the early 1970s. They serve with Aufklarungsgeschwader 51 at Bremgarten and AK52 at Leck. Essentially the reconnaissance systems are those of the RF-4C, but the basic airframe is that of the F-4E.

Apart from the MiG-25, which is apparently used for both tactical and strategic reconnaissance, the Soviet Union's main battlefield reconnaissance aircraft is the MiG-21R or Fishbed-H. This carries cameras and probably also IR linescan in a ventral pod and has small pods on the wingtips for electronic sensors.

Because tactical reconnaissance aircraft often operate in a high threat area, particularly when battlefield reconnaissance is required, unmanned remotely piloted vehicles (RPVs) or autonomously-guided drones have been developed for this task. They range from quite complex turbojet powered vehicles to expendable mini-RPVs which are little more advanced than radio-controlled model aircraft. The most widely used turbojet powered drone in NATO is the Canadair CL-89, which serves with the British, Canadian, West German, French and Italian armies. The CL-89 is launched from a ramp, follows a preprogrammed flight path and returns to friendly territory homing onto a beacon, where it parachutes back to earth. Either cameras or IR linescan can be carried.

During the war in Southeast Asia the USAF made extensive use of air-launched Teledyne Ryan AQM-34 reconnaissance drones, which were derived from the Firebee target drone. More than 20 variants were manufactured to undertake high, medium and low altitude missions by day and night. Guidance was either by radio command from an airborne or ground control post, or by a pre-programmed flight plan. The sensors were cameras or IR linescan and data could be transmitted to a ground or airborne relay station. The launch aircraft was a DC-130 variant of the Hercules, with recovery by parachute. Sometimes Sikorsky CH-3 helicopters retrieved the AQM-34 in the air after parachute deployment. This system is no longer in use, although it it believed that drones are in storage and could be returned to service quickly.

Rotary-winged drones have been evaluated as battlefield reconnaissance systems and as tethered platforms intended to raise sensors several hundred feet above their operating vehicles and relay visual or radar data to them. In West Germany Dornier developed the Do34 Kiebitz, which was fitted with a battlefield reconnaissance radar. Other rotary wing systems have been developed with low-light TV, still and TV cameras and IR sensors.

The basic reconnaissance tool remains the camera. For fast low-level work they need to use lenses of short focal length to give the widest angle of view. Often they are mounted in fans to give the maximum coverage. Higher flying aircraft can use a camera with a longer focal length lens to give a good image size. Various problems arise because of aircraft vibration and variations in temperature can also affect the quality of a photograph. Because the aircraft is moving during the exposure, a technique known as image movement compensation is used, which moves the film at the same speed as the object it is photographing so that the image will not be blurred.

The film used for photographic reconnaissance is usually black and white, but color film can be useful if this provides a greater contrast. A further alternative is color infra-red film, which is useful for penetrating camouflage. Of more use is the IR linescan sensor. This builds up a heat picture of the terrain, which can be recorded on tape or film for later analysis.

Photography and thermal imaging can be supplemented by radar reconnaissance. Sideways looking

An RC-135 ELINT aircraft (overleaf) approaches a tanker. The
RF-101 Voodoo (above) was an important tactical
reconnaissance aircraft in Vietnam, but is now out of service.
This OV-1B Mohawk (above right) carries a sideways-looking
airborne radar pod.

radar is especially effective because it makes use of
synthetic aperture techniques. This means in simple
terms that the distance an aircraft travels can be used as
though it were a very long radar aerial and so images of a
very high resolution are obtained.

The results are of little use, however, unless they are
accurately interpreted and the information is speedily
disseminated to the commanders who will make use of
it. These problems have long been recognized, but
solutions are not always easy to find. Early RF-4Cs were
fitted with film cassettes which could be ejected in flight
to users on the ground, but the system was found to be
unworkable and was discontinued.

The ideal system is one that provides the commander on the ground with 'real time' information, for
by the time an aircraft has returned to base and its film
has been processed the tactical situation may have
radically altered. The transmission of the SLAR
imagery, which is now possible, helps to achieve this
capability. SLAR is not defeated by cloud, haze and
smoke, but the images it produces, although they are
sharp, often cannot be identified because they are radar
blips and not a photographic image. The USAF has

used SLAR in conjunction with TEREC (tactical electronic reconnaissance) a system, which can identify a
target associated with a radio or radar emitter, as many
military targets will be. Clearly this is not a complete
answer. One promising development is the AN/UXD-1
system which is being tested by the USAF. This enables high quality photographs to be transmitted to a
ground station within ten seconds of them being exposed.

In spite of the development of advanced sensors and
stand-off capabilities, direct visual observation by a
reconnaissance crew is valuable in many situations.
One way to ensure that tactical targets are hit soon after
they are located is for the reconnaissance aircraft to
rendezvous with the strike force and lead it to the
target. The reconnaissance aircraft can then round off
its mission by photographing the results. Another possibility is that the reconnaissance crew can report their
visual observation by voice radio, provided that communications links are not jammed.

Where it is not essential that reconnaissance
imagery be received and processed rapidly (and this
applies especially to strategic reconnaissance), a number of techniques can be employed to gain the maximum information from the material. Hidden details
can be brought out during film processing. Computer
technology can be used to assist in interpretation of
photographic, IR and SLAR imagery, by automating
target searches and corellating the photo images with

map references. If the imagery is presented in digital form, then data processing can be used to sharpen images or alter contrasts to assist in identification and interpretation.

Because modern armed forces are so dependent upon radars and radio communications, electronic intelligence-gathering or ELINT is today as important as visual reconnaissance. ELINT aircraft operate on the borders of a potential enemy's territory scanning the wavelengths for electronic emissions. Every signal can be classified and identified, so that a picture can be built up of the opponent's air defenses, showing, for example, the positions of radars and control centers or the radio wavelengths used to communicate with enemy fighters. This information will be essential in planning one's own electronic-countermeasures, quite apart from the intelligence which can be gathered about enemy dispositions.

ELINT can make a valuable contribution to virtually every aspect of defense intelligence. ELINT aircraft have 'captured' the telemetry transmitted from Soviet missiles under test, so that the US intelligence agencies can accurately assess their performance. If the accuracy of Soviet missiles is known with some degree of certainty, then American planners can devise a basing system for their ICBMs which will defeat the opposition. In the absence of such intelligence, resources may be wasted on a needlessly sophisticated basing system. Alternatively deterrent forces may be

put at risk by presenting the adversary with an ICBM force he will be able to knock out.

At the tactical level, ELINT aircraft can enable a comprehensive library of enemy electronic systems to be built up. This means for example that a hostile warship using its search radar can not only be detected, but the radar itself can be identified and this will provide a clue to identifying the class of ship. The same information can be applied to the defense suppression mission, as every hostile radar emission will identify the SAM or AA gun which it is serving.

ELINT gathering is a highly classified activity and, although the aircraft types employed can be identified, the equipment they use and the way they operate is shrouded in secrecy. The USAF's ELINT aircraft are the Boeing RC-135s of the 55th Strategic Reconnaissance Wing at Offutt AFB, with aircraft operating on detached service to virtually all parts of the world. The RC-135 is basically a modified KC-135 Stratotanker. It has a range of over 5000 miles and is fitted with an airborne refueling receiver. The crew complement for the early RC-135C variant was quoted as 18, four of whom were the flight deck crew.

Ten versions of the RC-135 have been identified, each with a different combination of sensors, and indeed equipment variations occur within individual sub-types. The specific missions of these aircraft are not known, but each has its own permutation of SLAR and forward-looking radars, HF, VHF and UHF antennae,

The Bear-D version of the Soviet Tu-20 (left and above) is used for long-range reconnaissance and may also provide mid-course guidance for anti-shipping missiles.

plus a range of distinctive and mysterious aerials and radomes. The main sub-variants currently in service include the RC-135M with a thimble-shaped nose radome and antennae fairings on the rear fuselage; the RC-135U with SLAR and various fuselage-mounted antennae; the RC-135V combining features of the R and U and reportedly equipped to measure a radar's exact power output; and the RC-135W with SLAR and nose radomes.

The US Navy undertakes its own ELINT, which is primarily intended to gather data on Soviet naval electronic systems. Two fleet air reconnaissance squadrons carry out these missions; VQ-1 covers the Pacific from Agana on Guam, while the Atlantic and Mediterranean are the responsibility of VQ-2 at Rota in Spain. Both squadrons operate a mixture of EA-3B Skywarriors and EP-3E Orions, the former being able to operate from aircraft carriers. Small detachments of EA-3Bs are deployed aboard carriers for periods of up to several months. VQ-2 uses Athens in Greece, Sigonella on Sicily and Stuttgart, West Germany, on a regular basis, while VQ-1 operates a detachment from Atsugi in Japan. The EP-3Es can be used for overland ELINT missions, as well as for purely naval operations.

Lockheed EP-3Es are conversions of early-model Orion ASW aircraft and about a dozen are in service. They carry a 15-man mission crew, and can accommodate a complete relief crew for maximum efficiency on a 17 hour mission. An extensive range of detection and monitoring equipment is fitted, including an ALR-60 communications interception and recording system, an ALD-8 direction finding system, an ALQ-110 radar signal collection system and a frequency measuring receiver.

The Douglas EA-3B is a development of the A-3 Skywarrior attack aircraft, which entered US Navy Service in 1956. Although now obsolete in its original nuclear-strike role, the type continues to serve in small numbers in the tanker, reconnaissance and ECM roles, as well as for ELINT. The EA-3B carries a crew of seven, four more than the other variants. Power is provided by two Pratt & Whitney J57-P-10 turbojets, each developing 10,500lbs of thrust. Maximum speed is Mach 0.83 at 10,000ft, ceiling 41,000ft and range 2900 miles. Wingspan is 72ft 6in, length 76ft 4in and maximum takeoff weight 84,000lbs.

Two European NATO members operate small ELINT forces, using conversions of ASW patrol aircraft. The RAF has three Nimrod R Mk 1s in service for this task. They seve with No 51 Squadron, which is based at Wyton in the UK. The West German Navy has had five of its 20 Breguet Atlantic ASW aircraft converted for ELINT tanks. They are primarily concerned with Soviet naval activities in the Baltic and operate alongside the standard ASW Atlantics with Marinefliegergeschwader 3 'Graf Zeppelin' at Nordholz.

Electronic intelligence is greatly valued by the Soviet Union. Soviet aircraft seek this information on a routine basis and by monitoring NATO exercises. They also attempt to trigger a reaction from defense forces by provocative intrusions into national airspace, as they can learn much by monitoring the result. An ELINT version of the Antonov An-12 transport, the Cub-B, has been produced in addition to the ECM Cub-C. These aircraft may operate with Soviet Naval Aviation, as many Soviet ELINT sorties are concerned with naval operations. Another Soviet transport aircraft converted for ELINT work is the Ilyushin Il-18 Coot-A. It carries a large antennae fairing beneath the fuselage, as well as numerous radomes and aerials. Coot-A has operated around NATO central region and has probed the UK's air defenses.

Two older Soviet ELINT systems are converted bomber aircraft. They are probably less efficient than the modified transports, as they do not have the same amount of space for equipment and its operators. They are the Bear-E variant of the Tu-20 and the Badger-E and F versions of the Tu-16, both of which serve with Soviet Naval Aviation.

As with strategic reconnaissance, the introduction of ELINT or Ferret satellites has not made the aircraft systems unnecessary, but has rather provided a complementary source of intelligence. The satellites can cover areas deep in enemy territory, while the ELINT aircraft can cover such activities as naval exercises more effectively – if necessary by provoking a response from defensive systems.

The RAF's first VC 10 K Mk 2 in-flight refueling tanker, which was converted from a standard VC 10 airliner, pictured during testing.

7. TRANSPORT AIRCRAFT

Transport Aircraft

The modern transport aircraft has made it possible to airlift troops, their equipment and supplies over enormous distances at a speed hitherto unprecedented in warfare. Air transport operations can be strategic, involving the movement of forces between theaters of operations, or tactical airlift within the theater itself. A further aspect of tactical airlift is battlefield mobility, although this is more satisfactorily regarded as part of army aviation and is fully dealt with in the following chapter.

The advantages of air transport are obvious. Because of its speed of reaction, reinforcements can be rapidly lifted to a theater of operations from their peacetime stations, thus reducing the expense of permanent foreign garrisons, such as the US Army in Europe. Movement by air across the Atlantic is reckoned in hours, whereas by sea it would take days and even then the port of disembarkation could be far from the battlefield. However, air transport has definite limitations. The movement of large bodies of troops by air presents few problems, but without their weapons and equipment they are of little use. Even the massive C-5A Galaxy transport can only carry two main battle tanks and so clearly the airlift of an armored division from the USA to Europe would present problems.

One answer is the dual basing system for forces normally based in the United States, which would be needed in Europe in the event of war. These formations have equipment positioned both in the United States and in Europe, so that only the men need to be airlifted across the Atlantic. This is of course very expensive, as the cost of equipping these formations is doubled and the prepositioned equipment must be carefully stored and maintained.

During the American involvement in the Vietnam War no less than 98 percent of all supplies were transported by sea. This was despite strategic airlift operations on a massive scale. USAF Military Airlift Command's flying hours for 1967 alone were equivalent to 8750 flights around the world and the number of troops carried could have manned 85 US Army infantry divisions. The Vietnam statistics are probably misleading to some degree when applied to a war in Europe, because a massive amount of construction work was necessary in Southeast Asia and the materials inflated the sealift figures. Nevertheless, it indicates the basic problem of airlift operations, which are efficient in moving men, but unable to carry heavy equipment or gasoline and ammunition in sufficient quantity.

The US is especially concerned with the problems of mobility over large distances. Forces normally based in the USA may be required to reinforce NATO units in Europe or to bolster the US troops defending South Korea, while a further eventuality is the Rapid Deployment Force's movement to the Middle East, where no US Army or USAF forces are permanently stationed. All of these movements may have to be made

at short notice. In studying these contingencies, planners have had to balance the advantages and shortcomings of airlift, prepositioning of equipment and sealift. Air transport is fast and flexible, but its capacity is limited and its use is dependent on the availability of friendly airfields. Prepositioning, as well as being expensive, is inflexible, while sealift offers great capacity at the expense of speed and some flexibility. The planners' conclusions are that airlift will transport some 90 percent of the combat forces put into action during the first 30 days of the operation, with some equipment being prepositioned. In an extended conflict sealift will deliver between 90 and 95 percent of all supplies.

The availability of staging posts overseas may be of

The USAF's standard tactical transport aircraft is the C-130 Hercules. A C-130E (above) of the 36th Tactical Airlift Squadron overflies Mount St Helens. Strategic airlift is undertaken by the C-5A Galaxy and C-141 Starlifter (left). The C-5A has its nose visor raised. All USAF C-141As have been modified to C-141B standard (above right) by 'stretching' the fuselage by 23ft 4in.

crucial importance in airlift operations. Inflight refueling can go some way to make transport aircraft independent of the need for intermediate landings. Nonstop airlift operations across the Atlantic are quite feasible, of course, although in any case NATO controlled airfields are available on Iceland and the Azores. However, an airlift to the Middle East would require staging posts and supporting tanker aircraft would themselves require overseas bases. A related problem is the need to obtain permission to overfly foreign airspace. This has to be applied for through diplomatic channels well in advance and it could be refused. Thus air barriers can be created, which will require additional fuel to fly around with a consequent reduction in payload. These barriers can even make a proposed air route impassable. It has been found by the RAF that a peacetime flight from the UK to Hong Kong required at short notice can be more quickly mounted following a 'westabout' route over the USA, than the direct route over the Middle

East. In wartime the diplomatic complications of airlift operations could become even more difficult and time-consuming.

The United States strategic airlift capability comprises some 350 Lockheed C-5A Galaxy and Lockheed C-141B Starlifter transport aircraft of Military Airlift Command. In an emergency the Command can call upon approximately 460 civil passenger and cargo aircraft of the Civil Reserve Air Fleet (CRAF). These could carry 95 percent of military personnel requiring airlift, but only 35 percent of cargo. Cargo capabilities are to be increased by fitting wider doors and strengthened floors to civil transports in future. The capabilities of the regular forces are also being enhanced. The entire C-141A Starlifter fleet has been modified to C-141B standard, with fuselage lengthened by 23ft 4in and flight refueling capability added. The fuselage stretch gives a payload increase equivalent to 90 additional aircraft, without the need for further aircrews. The C-5A fleet is currently going through a wing-rebuilding program, which will extend aircraft life to 30,000 hours.

The USAF's strategic transports are flown by seven military airlift wings of the regular air force. These are supported by associate wings of the Air Force Reserve, which provide air and ground crews to supplement the regulars. In 1982 the military airlift wings were deployed as follows:

Unit	Base	Aircraft
60th MAW	Travis AFB, Ca	Two sqns of C-141B Two sqns of C-5A
62nd MAW	McChord AFB, Wa	Two sqns of C-141B
63rd MAW	Norton AFB, Ca	Three sqns of C-141B
436th MAW	Dover AFB, Del	Two sqns of C-5A
437th MAW	Charleston AFB, SC	Three sqns of C-141B
438th MAW	McGuire AFB, NJ	Three sqns of C-141B
443rd MAW	Altus AFB, Ok	One sqn of C-5A One sqn of C-141B

The C-5A Galaxy was designed from the outset to carry bulky items of military equipment. Its lower cargo deck can accommodate 16 light trucks, or 270 troops, while the upper deck can seat a further 75 troops. The C-5A is powered by four General Electric TF-39-GE-1 turbofans, which give a maximum speed of 571mph at 25,000ft. Range is 6500 miles when carrying maximum fuel plus an 80,000lbs payload. With a maximum takeoff weight of 764,500lbs, the Galaxy spans 222ft 8½in and is 247ft 10in long. Its runway requirements are comparatively modest for its size, the C-5A needing an 8000ft takeoff run and half that for landing. A flight crew of seven is normally carried.

The Lockheed C-141B Starlifter, 270 of which are in service, is the USAF's workhorse strategic transport

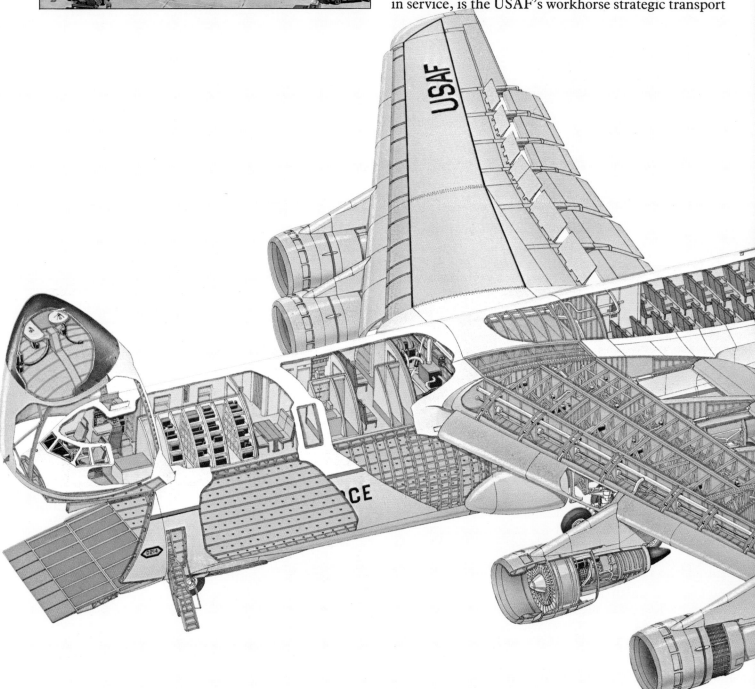

with good range and payload characteristics, the former enhanced by inflight refueling when necessary. The stretched aircraft is 168ft 4in long, with a wingspan of 159ft 11in and maximum takeoff weight of 324,900lbs. Power is provided by four Pratt & Whitney TF-33-P-7 turbofans of 21,000lbs thrust each. Cruising speed is 512mph and ceiling 45,000ft. With a 74,200lbs payload the Starlifter's range is 3200 miles. A flight crew of seven is carried and in addition to cargo the C-141B can carry troops or be converted to an aerial ambulance for medical evacuation.

The USAF's CX requirement calls for a new out-size-cargo transport which will be able to fly such equipment as tanks, self-propelled artillery and infantry fighting vehicles from bases in the USA direct

The USAF's C-5A Galaxy is Military Airlift Command's largest transport aircraft and is able to lift a maximum payload of over 220,000lbs (above left and below).

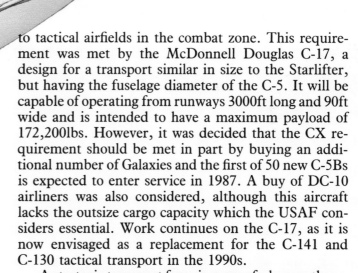

to tactical airfields in the combat zone. This requirement was met by the McDonnell Douglas C-17, a design for a transport similar in size to the Starlifter, but having the fuselage diameter of the C-5. It will be capable of operating from runways 3000ft long and 90ft wide and is intended to have a maximum payload of 172,200lbs. However, it was decided that the CX requirement should be met in part by buying an additional number of Galaxies and the first of 50 new C-5Bs is expected to enter service in 1987. A buy of DC-10 airliners was also considered, although this aircraft lacks the outsize cargo capacity which the USAF considers essential. Work continues on the C-17, as it is now envisaged as a replacement for the C-141 and C-130 tactical transport in the 1990s.

A strategic transport force is more of a luxury than a necessity for the Soviet Union, which does not have to move its forces over large areas of ocean to reach the probable battlefields of World War 3. However, the vast extent of the Soviet homeland makes air transport within its frontiers very useful. With expanding Soviet influence in such troubled areas as the Middle East, Africa, Southeast Asia and Central America, a long-range airlift capacity has become necessary to support Soviet client states and associates in distant regions. Nevertheless, Military Transport Aviation remains

primarily concerned with supporting Soviet ground forces and with transporting the airborne divisions. At the beginning of 1982 Military Transport Aviation had a strength of some 600 medium- and long-range transport aircraft, more than 180 of which were long-range Ilyushin Il-76 Candids and Antonov An-22 Cocks. In addition the Soviet Union can make use of the total resources of Aeroflot, which would add about 1300 long- and medium-range passenger and cargo aircraft to the military transport fleet. Aeroflot could also provide several thousand short-range transports and helicopters.

The Ilyushin Il-76 is really a successor to the Antonov An-12 Cub tactical transport, but because it has twice the payload and range it can be used for strategic airlift. This was demonstrated when weapons were sent to Ethiopia in late 1977. Although similar in size to the C-141B Starlifter, the Il-76 is more powerful. It is also able to use rough tactical airstrips, unlike the C-141, and can operate comfortably from a 5000ft runway. The Il-76 has a maximum payload of 88,200lbs and can lift this over a distance of more than 3000 miles. Maximum range is over 4000 miles, but unlike the US transports the Il-76 cannot be refueled in flight. Power is provided by four 26,450lbs Soloviev D-30KP turbofans, which give a maximum cruising speed of 466mph. Wing span is 165ft 8in and length 152ft 10in, with a maximum takeoff weight of 375,000lbs. A flight crew

of six is carried and many aircraft carry a defensive armament of two radar-directed 23mm cannon in a manned tail turret.

The massive Antonov An-22 Cock is the Soviet Union's heavylift strategic transport, with some 50 in military service and an equal number with Aeroflot. It can accommodate main battle tanks, or the tracked launch vehicle for twin SA-4 surface-to-air missiles, as part of its 175,000lbs maximum payload. The An-22 is turboprop powered, with four 15,000shp Kuznetsov NK-12MV engines driving massive 20ft diameter contra-rotating propellers. These give a maximum speed of 460mph, with a service ceiling of 25,000ft. Range with a payload of 100,000lbs is 6800 miles. With a maximum takeoff weight of 550,000lbs, the An-22 spans 211ft 4in and is 189ft 7in long. Surprisingly in view of its bulk, the An-22 has been designed to operate from rough, unprepared airstrips. This reflects the Soviet concern with tactical airlift in support of ground forces, with strategic airlift being only secondary. In World War 3 the Soviet strategic transport fleet is likely to be used for the rapid movement of forces from outlying military districts to the battlefronts. The An-22 could also be used to fly heavy equipment into a bridgehead seized by the airborne divisions.

Tactical airlift is concerned primarily with the logistic support of the fighting armies. Urgently-needed supplies and reinforcements can be more

The Soviet long-range, heavy-lift transport aircraft is the An-22 Cock (top left). It has a maximum payload of some 175,000lbs. The turbofan-powered Il-76 Candid (above left) has a range of 4,000 miles. A USAF C-130E parachutes its cargo (above), which comprises up to 16 one ton loads.

rapidly carried to the front by air than by land transport. Airborne forces operating behind enemy lines, or pockets of resistance surrounded by an advancing enemy, can only be resupplied by air and if necessary they can be evacuated by the same means. Tactical transports will provide an essential link between the strategic airlift of forces into the theater of war and the battlefront, by ferrying troops from the secure 'airheads' in rear areas to tactical airstrips in the combat zone.

The USAF's tactical airlift commitment is carried out by 14 squadrons of Lockheed C-130 Hercules transports of Military Airlift Command. Until 1974 these assets were controlled by Tactical Air Command, reflecting the close co-operation between the tactical airlift squadrons and the ground forces they support. However, the change in command represented a rationalization of resources and a recognition that the Hercules' 4600-miles maximum range could make it a useful supplement to the strategic transport force in an emergency. MAC's regular peacetime strength can be doubled by the 51,000 reservists of the ANG and

AFRES trained in air transport duties. Over 300 tactical transport aircraft, including C-130s and the older Fairchild C-123 Providers and the DHC C-7 Caribou, are in service with the ANG and AFRES.

The current deployment of MAC's tactical transport units is as follows:

Unit	Base	Equipment
314th TAW	Little Rock AFB, Ar	Four sqns of C-130E/H
316th TAG	Yokota AB, Japan	One sqn of C-130E
317th TAW	Pope AFB, NC	Three sqns of C-130E
374th TAW	Clark AB, Philippines	One sqn of C-130E
435th TAW	Rhein-Main AB, W. Germany	One sqn of C-130E
463rd TAW	Dyess AFB, Texas	Three sqns of C-130H
616th MAG	Elmendorf AFB, Alaska	One sqn of C-130E

MAG = military airlift group; TAG = tactical airlift group; TAW = tactical airlift wing.

The Lockheed C-130 Hercules entered USAF service in December 1956 and over 1000 have been built for the American services alone. The aircraft has been exported to over 50 countries, including among the NATO allies Belgium (12), Canada (33), Denmark (3), Greece (12), Italy (14), Norway (6), Portugal (5), Spain

(12), Turkey (8) and the United Kingdom (66). The C-130E version is powered by four 4050shp Allison T-56A-7 turboprops, which give a maximum cruising speed of 368mph. Maximum payload is 45,000lbs, which can be carried over a range of 2400 miles. Take-off weight is 175,000lbs and dimensions are 132ft 7in wing span and 97ft 9in length. The C-130H, an improved version with more powerful engines, remains in production. Deliveries of new C-130H aircraft have been made direct to Air National Guard squadrons, as part of the modernization of the reserves begun in the 1970s.

In 1972 steps were taken to find a successor to the veteran C-130. The Advanced Medium STOL Transport requirement produced two designs (the YC-14 and YC-15) which were tested by the USAF. However, funding was deleted from the 1979 defense budget and MAC will have to await the introduction of the C-17 in the 1990s before it can begin replacing its C-130 fleet.

Among the European NATO members the UK and France have the largest transport forces, partly as a result of their continuing commitments in former colonial territories. The RAF has four tactical transport squadrons equipped with the Hercules and based at Lyneham and a single strategic transport squadron flying 4200 miles range VC 10 transports from Brize Norton. Thirty of the standard Hercules C Mk 1 are being 'stretched' by an extra 13ft 4in fuselage section. Redesignated Hercules C Mk 3, they have a 40 percent increase in cargo capacity and 92 paratroops can be lifted instead of the C Mk 1's 64. These forces are primarily committed to NATO, but they also support operations outside the Alliance's area. France's tactical transport force flies the Franco-German C.160 Transall, which equips Escadre de Transport 61 at Orleans-Bricy and is replacing the elderly Noratlas with ET64 at Evreux. The Transall C.160 also equips three transport wings of the Luftwaffe. Powered by two Rolls-Royce Tyne Mk 22 turboprops of 6100shp, it has a range of 2830 miles with a 17,650lbs payload.

The An-12 Cub remains the mainstay of Soviet Military Transport Aviation, with over 400 in service. Most are based in the western military districts of the USSR. Powered by four 4000shp Ivchenko AI-20K turboprops, the An-12 has a maximum cruising speed of 400mph. With a 22,000lbs payload, the An-12's range is 2100 miles. Maximum payload is 44,000lbs and up to 100 troops can be accommodated. However, because the cargo cabin is unpressurized, it has to fly at low altitudes when carrying troops, thus reducing range. Maximum takeoff weight is 134,500lbs and dimensions include a span of 124ft 7in and a length of 108ft 6in. Twin 23mm cannon are carried in a manned tail turret.

Various methods can be used by tactical transports to deliver supplies to forward areas. The preferred method is to land and unload, as this avoids the complications and expense of parachute systems and allows the maximum load to be delivered. The landing fields used are likely to be primitive 3500ft dirt airstrips, constructed with the minimum of labor. They can present problems to the modern transport pilot, used to operating from 10,000ft concrete runways. For example a turboprop using reverse thrust can create a cloud of dust over the entire landing area. The problem is exacerbated at night, when airstrip lighting will be considerably less efficient than that of a permanent airfield. Even without dust clouds, an unsurfaced airstrip will tend to blend into the surrounding landscape. Yet all these problems are comparatively insignificant when compared with the dangers of enemy artillery or AA fire, or air attack. For this reason, the aircraft will keep their time on the ground to a minimum. After landing the transport must taxi off the strip to clear it for following aircraft, it will unload with engines running and take off again as soon as possible. If the approaches to the runway are threatened by enemy fire, the transport aircraft will make a very steep let-down to land, keeping out of range of small arms fire for as long as possible. The procedure is named the Khe Sanh approach, after the besieged outpost in Vietnam where the technique was first applied.

If conditions on the ground are too dangerous to allow the transport to land, supplies can be parachuted. The USAF's Low-Altitude Parachute Extraction System (LAPES) used with the C-130 requires the pilot to overfly the airstrip at a speed of 130 knots and a height of only five feet. A drogue parachute, attached to a special sled-like cargo pallet, is trailed from the rear cargo hatch. As the canopy deploys, it pulls the pallet and its cargo out of the aircraft and onto the runway where it skids to a halt. This method is accurate and fast, but it can be very dangerous for personnel on the airstrip and the low-flying aircrew must be careful to avoid obstacles on the ground.

A similar system to LAPES, the Ground Proximity Extraction System (GPES), was safer. It made use of an arrestor system on the airstrip, which engaged a hook attached to the cargo pallet after the aircraft touched down. The pallet was pulled out and the aircraft meanwhile accelerated and took off. The real disadvantage of GPES was that because it required special ground equipment, it lacked the flexibility of LAPES and so only a few sets of the system were produced.

If the landing zone is obscured by cloud or mist, parachute drops can be made from medium altitude guided by ground radar. However, very precise flying is required of the transport's crew if the loads are not to be hopelessly scattered. Three tactical airlift squadrons in MAC are equipped with the adverse weather aerial delivery system (AWADS). This comprises a very accurate ground-mapping radar carried by the C-130 and linked to the aircraft's flight control instruments. AWADS is thus independent of ground radars, which even if they are available in the right place, are vulnerable to enemy action. It is also more accurate than

The twin-turboprop Transall C.160 (right) is flown by the air forces of France and West Germany. This view of the VC 10 K Mk2 (below) shows its underwing refueling pods and the third refueling point beneath the rear fuselage. The An-12 Cub (bottom right) is being replaced by the Il-76 as the standard Soviet tactical transport, but over 400 remain in service.

ground-controlled drops.

The USAF's parachute container delivery system is in fact a one ton load of fuel, food or ammunition lashed to a pallet and then covered by a shroud. The C-130 carries up to 16 of these, which are placed on rollers on the floor of the cargo compartment. Over the dropping zone the pilot opens the rear loading doors and lowers the cargo ramp, while the loadmaster releases the restraining lashings. The pilot then pulls up the aircraft's nose, allowing the pallets to roll out gently and be parachuted.

Evacuation of the wounded from the battle zone to permanent hospitals is an important task for military transport aircraft. Just as the helicopter enables the wounded to be removed from the battlefield to medical care as soon as possible, so aeromedical airlift operations can clear the wounded from field hospitals. This not only ensures better attention for the wounded at permanent and well-equipped hospitals remote from the war zone, but it also relieves the burden of caring for them from the overworked medical teams in the field.

The USAF operates a special unit for this role, the 375th Aeromedical Airlift Wing based at Scott AFB, Illinois. It flies the C-9A Nightingale, a version of the DC-9 airliner, specially-equipped and carrying doctors and nurses to care for the patients in flight. Detachments of the unit are based at Rhein-Main in West Germany and Clark AB in the Philippines. Standard

transport aircraft can also be used in this role, including the C-141 Starlifter. During the Vietnam War Lockheed C-141As flew 6000 such missions between 1965 and 1972.

Army airborne forces are of course a priority claimant for military airlift resources. However, they make great demands on what will undoubtedly be a scarce asset in wartime. During the US exercise 'Bright Star' in 1981 it was found that it would take 10-14 days to deploy one of the US Army's two airborne divisions to Egypt, although an airborne brigade could be rushed to the area in 48 hours. This gives some idea of the magnitude of the problem of using airborne forces on a large scale. A US light infantry division, which could be used as an air landing force rather than a true airborne formation, would require 1230 C-5A and C-141B sorties to transport to Europe. With the USAF transport fleet presently available this would take several weeks to accomplish.

A more realistic approach would be to use smaller units to reinforce troops already in the theater of operations at short notice. For example, a 600-man paratroop force with its equipment can be airlifted to Europe from the USA in ten C-141B Starlifters. This would involve an eight hour flight with a drop into the battle area. The UK may be better placed to employ its airborne troops in this manner, but it will require a force of 21 Hercules C Mk 1s to lift a battalion to its drop zone. One weak-

A C-7A Caribou transport (above left) demonstrates the technique of low-altitude parachute extraction. One of the most valuable features of the C-5A Galaxy is its ability to transport bulky equipment, such as this Roland fire unit (above).

ness of airborne troops is their lack of heavy equipment, such as main battle tanks or artillery, and this will put them at a disadvantage when fighting armored and infantry divisions. Consequently the employment of airborne troops must be carefully co-ordinated with that of friendly ground forces.

The Soviet Union employs airborne troops on a large scale and currently maintains seven airborne divisions. They are well equipped with BMD amphibious assault vehicles, giving them better mobility and firepower than the paratroops of the NATO nations. The Soviet Union is therefore equipped to undertake large-scale airborne assaults, using paratroops, reinforced with air landing forces once airfields have been secured. As Military Transport Aviation does not have MAC's transoceanic reinforcement role, Soviet airborne troops are likely to enjoy a higher priority for airlift resources than their US counterparts. In addition to the airborne divisions, the Soviet Union has a number of air assault brigades, which can be used for small scale assaults on key objectives, such as nuclear weapon storage sites.

Raids against high value targets behind enemy lines will be made on a much smaller scale by highly-trained Special Forces troops, who will require air support to infiltrate enemy territory, to supply them during their operations and finally to evacuate them. The USAF maintains three squadrons trained and equipped primarily to support Special Forces operations. They are the 1st Special Operations Squadron (SOS) at Clark AFB in the Philippines, the 7th SOS at Rhein-Main in West Germany and the 8th SOS at Hurlburt Field, Florida, all flying the MC-130E version of the Hercules. Code-named Combat Talon, the mission of these aircraft is the covert infiltration and recovery of Special Forces from enemy territory. They operate at ultra-low level at night, using terrain masking for concealment. Special navigation equipment is fitted. The troops and their equipment are dropped by parachute and evacuation can be made using the STAR (surface-to-air recovery) system. This apparatus consists of a pair of caliper-like arms mounted on the aircraft's nose. The soldier on the ground wears a special harness attached to a line which is raised by a helium-filled balloon. This is snatched by the aircraft's nose apparatus and the soldier is winched up into the aircraft.

The helicopter is also very useful in special operations. The USAF's 1st Special Operations Wing at Hurlburt Field, which experiments in and teaches the air aspects of unconventional warfare, flies the Sikorsky

CH-3E, the Bell UH-1N and Sikorsky HH-53. The latter helicopter has been extensively used on special operations, as well as in its primary role of combat rescue. During the Vietnam War, in November 1970, HH-53s carried a force of raiders to the Son Tay prison camp, 23 miles from Saigon. Their mission was to release American PoWs. The plan was boldly conceived and brilliantly executed, but intelligence had failed to discover that all the prisoners had been removed to other camps. In May 1975 HH-53s and the similar CH-53 (intended for special operations, but lacking the rescue helicopter's inflight refueling capability) participated in the operation which led to the release of the crew of SS *Mayaguez*, an American freighter illegally seized by Cambodian gunboats.

These successful air operations should be set against the much publicized failure to release the US hostages from Tehran in April 1980. The helicopters employed in this mission were RH-53s – a similar design to the HH-53 but with the primary mission of minesweeping. Operating from the carrier USS *Nimitz*, three of the eight machines became unserviceable before reaching the desert landing strip which was to be the jumping-off place for the rescue mission. This led to the abandonment of the mission, which required six helicopters to transport the Special Forces troops to Tehran. The failure was exacerbated when one of the helicopters collided with a supporting Hercules transport taking off from the desert strip.

The Iranian debacle was due to a failure in planning and sheer misfortune, rather than to any fundamental weakness in equipment. It is very likely that helicopter operations will support clandestine missions in Europe during World War 3. CH-53s are based at Sembach in West Germany, ostensibly to transport mobile radars, but it is probable that they have a secondary special operations role. Soviet special forces troops controlled by the KGB and GRU (military intelligence) will operate extensively behind NATO lines in World War III. They will be supported by transport aircraft of Military Transport Aviation and by the assault helicopters of Frontal Aviation.

Air refueling operations are one of the most valuable of air force support functions. Tanker aircraft are 'force multipliers', enabling tactical aircraft to operate at extended ranges, or for longer periods and with a heavier warload. Their contribution to strategic bomber operations is vital and indeed all USAF tanker aircraft are controlled by SAC. Tanker aircraft will enable the NATO air forces in Europe to be rapidly reinforced with tactical fighter squadrons from the USA and Canada in the event of war. However, because air refueling is such a useful adjunct to a variety of combat and support missions, tanker aircraft will be very scarce assets in war. Their allocation will be carefully controlled and it is unlikely that tanker support will be available for every mission that could benefit from it.

At present the USAF's aerial refueling force comprises 615 Boeing KC-135 tanker aircraft, including 128 assigned to the Air National Guard and AFRES. All these aircraft are controlled by SAC and have the primary mission of supporting strategic air operations. However, a porportion of the tanker force can be detailed to support tactical air operations. All strategic bomber and C^3 aircraft are capable of being refueled in flight, as are the great majority of tactical fighter aircraft, reconnaissance, airborne control aircraft, strategic transports and the tankers themselves. SAC is currently upgrading its capabilities by introducing the new KC-10 Extender tanker/transport, which is especially valuable for supporting the overseas deployment of tactical fighters, and by extending the service life and efficiency of the KC-135s by a re-engining program.

The tanker aircraft are either deployed as part of SAC bombardment wings, as detailed in Chapter One, or as separate air refueling units. The deployment of these forces at the beginning of 1982 was as follows:

Unit	Base	Composition
6th Strategic Wing	Eielson AFB, Alaska	KC-135As On TDY
11th Strategic Group	Fairford, UK	KC-135As on TDY
34th Strategic Squadron	Zaragoza, Spain	KC-135As on TDY
100th Air Refueling Wing	Beale AFB, Ca.	Two sqns of KC-135A/Q
305th Air Refueling Wing	Grissom AFB, In.	Two sqns of KC-135A
306th Strategic Wing	Mildenhall, UK	TDY KC-135A/Q
340th Air Refueling Group	Altus AFB, Ok.	One sqn of KC-135A
376th Strategic Wing	Kadena AB, Okinawa	One sqn of KC-135A
384th Air Refueling Wing	McConnell AFB, Ks.	Two sqns of KC-135A

TDY = temporary duty (ie detached from US-based units).

The KC-135 Stratotanker family bears a striking resemblance to the Boeing 707 airliners and in fact the military and civil aircraft are parallel developments, with a common ancestor in the Boeing Model 367-80 of

An MC-130E releases flares over its drop zone, closely followed by a second aircraft. The MC-130E's mission is infiltration, supply and recovery of Special Forces troops behind enemy lines.

1954. Powered by four Pratt & Whitney J57-P-59W turbojets each developing 13,750lbs of thrust with water-injection at takeoff, the KC-135A has a cruising speed of around 530mph at 35,000ft. Maximum takeoff weight is 316,000lbs, wingspan is 130ft 10in and length 134ft 6in. Total fuel capacity is 31,200 gallons, housed in integral wing tanks, underfloor fuselage tanks and a rear fuselage tank. The KC-135A can fly out to a radius of 3000 nautical miles to offload 24,000lbs of fuel, or it can supply 120,000lbs of fuel at a 1000nm radius. As many KC-135s are fitted with an air refueling receiver system these operating radii can be extended by the tankers themselves being refueled in flight. Operating at maximum weight the KC-135A will require nearly 14,000ft of runway for takeoff.

The KC-135A is operated by a crew of four, comprising two pilots, a navigator and a boom operator. The cabin can be used for personnel or cargo transport, with accommodation for up to 160 troops, or 83,000lbs of cargo. This is useful for overseas deployment of tactical squadrons, allowing maintenance personnel and ground equipment to be carried. Fuel is transferred by means of a boom fitted beneath the rear fuselage. This is lowered, telescopically extended and then maneuvered by the boom operator until it connects with the refueling receptacle in the receiver aircraft. Fuel can then be transferred at the rate of 5850lbs per minute. Two rows of lights are mounted beneath the forward fuselage to provide the receiver aircraft with aid in lining up with the tanker.

The KC-135Q version of the Stratotanker (56 produced by modifying KC-135A airframes) provides tanker support for the SR-71 Blackbirds and has a special fuel system capable of handling the JP-7 fuel used by the aircraft and TACAN equipment to facilitate aerial rendezvous. Apart from the exotic SR-71 all USAF units assigned to NATO are converting to JP-8 fuel from JP-4. The new fuel is to be standardized for all NATO aircraft, thus contributing to interoperability and also to safety, as it has a higher flashpoint than JP-4. Another development that will contribute to compatibility between the NATO air arms is the fitting of a hose drum unit to the belly of KC-135As. This means that aircraft of the US Navy and of the NATO allies, who all use the probe-and-drogue refueling system, can refuel from USAF KC-135s without prior notice. Without this installation, a drogue unit has to be fitted to the boom before takeoff if probe-equipped aircraft are to be refueled.

The most significant modification in prospect for the KC-135A is its re-engining with license-built versions of the French SNECMA CFM-56 turbofan giving 60 percent more power for 25 percent less fuel. Plans call for 344 of the existing fleet to be so modified by 1988, receiving the new designation KC-135R. The advantages of the program are that takeoff performance will improve, engine noise levels will be reduced, and most significantly, because the tanker's engines will consume less fuel, it will have more available to pass on. It is estimated that the KC-135R will be able to offload up to twice as much fuel as it predecessor and its improve takeoff characteristics will allow it to operate from four times as many bases.

A USAF KC-135A Stratotanker (above) refuels the first C-141B conversion during development trails. An RAF Victor K Mk 2 tanker streams its three hose and drogue refueling units (below). The USAF's KC-10A Extender (below left) combines the duties of cargo transport and tanker aircraft.

The KC-135 tanker will soldier on into the 21st century, but it will be supplemented by up to 60 McDonnell Douglas KC-10A Extenders, a tanker/cargo version of the civil DC-10. By May 1982 the USAF Reserve had 17 of these in service with the 78th Air Refueling Squadron at Barksdale AFB, Louisiana, with a second ARS due to form during the year. The KC-10A is intended to reduce the US forces' reliance on overseas staging posts – a weakness demonstrated during the 1973 Arab-Israeli War when many countries denied landing rights to USAF supply flights bound for Israel. The Extender will be able almost to double the range of a fully-loaded C-5 Galaxy, or accompany an overseas fighter deployment, refueling the aircraft and carrying maintenance men and their equipment.

The KC-10A is powered by three 52,500lbs General Electric CF6-50C2 turbofans, which give it a maximum cruising speed of 595mph. Range with maximum cargo carried is 4370 miles and at a maximum takeoff weight of 590,000lbs an 11,000ft runway is required. Fuel load is 117,500lbs in underfloor tanks in the fuselage, plus 238,565lbs in wing tanks. This can be transferred using either a boom, or a hose and reel unit, both of which are fitted on all aircraft. The cabin can accommodate up to 80 passengers, plus a crew of five, or up to 169,400lbs of cargo. Operating out to a radius of 1900nm, the KC-10A can offload 200,00lbs of fuel.

The US Navy and Marines, in common with several NATO air forces, use the probe-and-drogue method of aerial refueling. This involves the tanker aircraft trailing a hose with a drogue on the end. The receiver aircraft then maneuvers to engage the drogue with the

An HH-53B Super Jolly Green Giant rescue helicopter (top) maintains a low hover while winching a rescued airman aboard. In order to improve the helicopter's night rescue capability, this HH-53 is fitted with Pave Low III sensors (above).

refueling probe attached to his aircraft. Once the connection is made fuel is transferred. The KA-6D modification of the Intruder is used as a shipborne tanker aircraft while the Marines use the KC-130F and R. Attack aircraft also use the buddy refueling technique. This involves aircraft flying in pairs, one fuel-laden aircraft carrying a hose and drogue unit in a refueling pod, which is used to transfer fuel to the second attack aircraft to extend its range. This technique is also used by several NATO air forces, for example the RAF's maritime attack Buccaneers and the Marineflieger's Tornados.

In Europe the RAF maintains a tanker force comprising two squadrons of Victor K Mk 2 tankers (16 aircraft). These converted strategic bombers are fitted with a three-point hose and drogue refueling system. The Falklands Crisis revealed the inadequacy of so small a force and a small number of Vulcan bombers and Hercules transports have consequently been converted to tanker aircraft. This is a stopgap measure pending the availability of VC 10 tanker aircraft, which are due in service in the mid-1980s to supplement the Victors. Nine Super and Standard VC 10 airliners are undergoing conversion. France's Armée de l'Air maintains a single squadron of C-135F tankers, which are similar to the USAF's KC-135As, but equipped for probe-and-drogue refueling. Additionally 15 of the new production C.160 Transall transports will have provision for a hose-and-drogue unit to be fitted, ten aircraft actually being thus equipped initially.

An interesting answer to the shortage of tanker aircraft was put forward by the Boeing Company in 1981. It was suggested that a civil tanker reserve force be created by converting airliners as tankers. British Airways' Boeing 757s due to be delivered in 1983 could be converted at a cost of £44 million. More ambitiously, a NATO reserve tanker force could be created by modifying the Boeing 737s flown by the airlines of European NATO powers. The disadvantage of such schemes, apart from finding the money from slender defense budgets, is in finding the time to train civil crews up to the necessary level of competence in a specialized branch of military aviation.

At present the Soviet Union lags behind NATO in provision of inflight refueling for tactical aircraft. Some 30 modified M-4 Bison bombers serve with Long Range Aviation and a further 70 Tu-16 Badgets provide Naval Aviation with tanker support. Whether the introduction of the Ilyushin Il-76 tanker version reported to be under development will increase Soviet use of this support service remains to be seen. Unlike the United States, the Soviet Union has no worries about intermediate landing facilities when moving her forces within the Soviet block. However, her increasing involvement in the Third World may act as a spur to developing a larger tanker force. The Soviet air force (in common with the USAF and RAF) is reportedly interested in acquiring a large subsonic aircraft which could serve in transport, maritime reconnaissance, cruise-missile carrying and tanker roles.

Combat rescue is a valuable activity quite apart from its humanitarian aspects and the beneficial effect that an efficient rescue organization will have on aircrew morale. Military aircrew represent a considerable investment in costly and time-consuming training and experience. Therefore any effort made to recover aircrew from enemy territory or the sea is sound military policy, as much as an act of humanity. The United States maintains a considerable force of rescue aircraft divided between the US Coast Guard, concerned primarily with air/sea rescue around the American coasts and the USAF's Aerospace Rescue and Recovery Service, (ARRS) which is part of Military Airlift Command. The Coast Guard, which will be embodied into the US Navy in time of war, operates C-130 Hercules for long-range search and rescue (SAR) and Sikorsky HH-52A and HH-3F helicopters.

ARRS has a combat rescue role, extensively practiced during the war in Southeast Asia where 3883 lives were saved. The aircraft used then, the HC-130 variants of the Hercules and the HH-3 'Jolly Green Giant' and HH-53 remain in service. It is a moot point whether the successes obtained over the jungles of Southeast Asia could be repeated in the hostile environment of Central Europe or even over the deserts of the Middle East. Nevertheless, the USAF maintains six squadrons ostensibly for this task, although two US-based units are primarily concerned with supporting SAC's ICBM sites. The HC-130 Hercules are equipped with spacecraft reentry tracking systems, search radars for ocean surveillance and inflight refueling equipment to extend the range of their accompanying helicopters.

The HH-3E is a twin-engined, all-weather rescue helicopter, powered by two 1400shp General Electric T58-GE-10 turboshafts. Rotor diameter is 62ft, fuselage length 54ft 9in and maximum loaded weight 21,500lbs. The HH-3E has a maximum speed of 166mph and a range of 625 miles, which can be extended by inflight refueling. It is fitted with a rescue hoist, which has a 240ft cable able to lift loads of 600lbs. The later HH-53 is a derivative of the US Marine CH-53 heavy assault helicopter. It is armed with three 7.62mm Miniguns each with a maximum rate of fire of 4000rpm, which are used to suppress hostile groundfire during rescue missions. In addition it is extensively armored. Maximum weight is 42,000lbs and a crew of up to eight is carried – pilot, co-pilot, flight mechanic and five pararescuemen to assist survivors and man the armament. The HH-53's 600-mile range can be extended by flight refueling.

The HH-53s are scheduled to be replaced by HH-60D Night Hawk variants of the US Army's UH-60A Black Hawk, with the older helicopters reverting to the troop transport/special operations role. The HH-60D will operate at night and at low level, using terrain-following radar and FLIR in an attempt to evade enemy defenses during its rescue missions. Should this fail, the helicopter is armed with three Miniguns and may be given Stinger anti-aircraft missiles and Hellfire antitank missiles. It is equipped for inflight refueling. An interim model, the UH-60E, is likely to be used by Special Forces units to gain experience in operating the helicopter. The USAF's experience in combat rescue is unrivalled, most nations merely having a US Coast Guard-style search and rescue organization, usually equipped with helicopters.

The AH-64 Apache is the US Army's latest attack helicopter.
Armed with Hellfire advanced anti-tank missiles it can fight in all
weathers by day or night.

8. ARMY AVIATION

Army Aviation

The helicopter will be a major participant in the land battles of World War 3. The US Army is the world's largest user of helicopters, with 9000 of its 10,000 aircraft being rotary-wing machines. A wide range of Army helicopter roles has been developed to the point where virtually all land operations make use of helicopter support.

These roles encompass battlefield mobility for infantry, resupply of ammunition and fuel, and heavy lift for engineer and artillery units. The helicopter has been used for casualty evacuation since the Korean War. Its use for scouting and surveillance is almost as old, whereas the electronic warfare helicopter is a much more recent development. However, it is the armed helicopter, with its ability to attack enemy troops and more significantly armor, which has made the greatest impact on modern tactics. Primarily because of its importance in the antitank role, the helicopter has itself become the target of enemy air forces; attack helicopters and close air support aircraft will both operate in the anti-helicopter role over the battlefield.

Most armed forces have come to broadly the same conclusions about the demarcation between army aviation and air force responsibilities. Army air units supply battlefield mobility and undertake related support functions, while the air forces control close air support aircraft (except the armed attack helicopter which the Army controls) and tactical airlift. In 1967 after a bitter wrangle the US Army passed over its twin-engined C-7 Caribou transport aircraft to the USAF, leaving light transport and liaison aircraft to the Army. Thereafter the US Army and USAF followed the areas of responsibility outlined above. There are a number of exceptions to the general rule however. For example in the UK the RAF controls troop transport and heavy lift helicopters, while the Army Air Corps flies lighter reconnaissance and antitank helicopters. At present all Soviet battlefield helicopters are controlled by the air force.

The most serious criticism levelled at the army helicopter is its vulnerability to groundfire. The statistics for Vietnam, the first conflict in which helicopters were used on a large scale, would seem to bear this out, with over 16,000 helicopters brought down by enemy fire or in accidents. Yet to put this figure into perspective, it should be remembered that helicopters were used in great numbers at a very high sortie rate. During the Lam Son 719 operation into Laos in February/April 1971 helicopter losses were 107 machines, but this was only one helicopter for every 4000 sorties flown.

It would nonetheless be futile to deny that helicopters will suffer a high loss rate in battle. This can be alleviated in several ways. Making use of terrain masking and other cover and co-ordinating helicopter actions with artillery fire and close air support sorties will help. Improvements in design make the latest generation of helicopters less vulnerable to groundfire,

by the use of armor, self-sealing fuel tanks and 'redundant' structures and controls. Redundancy means designing an airframe in such a way that if a primary load-bearing member is hit, others will take up the load and the structure will not fail. The same philosophy will result in duplicated control runs, as widely separated as possible in the aircraft, so that if one is severed the second will take over its function. Finally, helicopters brought down are by no means invariably complete write-offs. It is therefore possible to recover them, effect quick repairs and return them to service. In Vietnam around 80 percent of downed helicopters were recovered and about one-third of these could be returned to service in 24 hours. It is unlikely that such a high recovery rate would be possible over the European battlefield.

The standard US Army troop transport helicopter is the Bell UH-1 'Huey', with about 4000 in service. Most of these are the UH-1H variant, which is similar to the UH-1Ds used in Vietnam except for its more powerful engine. Most US Army divisions have a combat aviation battalion and infantry divisions have several troop lift helicopter companies with a strength of between 20 and 25 UH-1s. Air Cavalry units rely exclusively on helicopter transport. The UH-1 carries two pilots plus a squad of infantry. The UH-1H is powered by a 1400shp Lycoming T53-L-13 turboshaft and has a maximum speed of 127mph at maximum weight. Range is 318 miles and payload is nearly 4000lbs. Maximum takeoff weight is 9500lbs and dimensions include a rotor diameter of 48ft and a fuselage length of 41ft 1in. The 220cu ft volume cabin accommodates up to 14 troops, two of whom may be gunners manning 7.62mm M60 machine guns mounted by the doors. The UH-1H can also be fitted with mine dispensers and used to lay minefields more rapidly and safely than ground vehicles.

The stalwart UH-1s are in the process of being supplemented by the Sikorsky UH-60A Black Hawk, with the planned total buy of the new helicopter standing at 1107 machines. However, the US Army will continue to rely heavily on the UH-1Hs until the end of the century. By mid 1982 300 UH-60As had been delivered to the US Army, its intended roles being troop transport and medical evacuation. This helicopter has been designed to operate with the minimum of attention from groundcrew in the field, under one hour of maintenance being required for one hour of flight during the first year of service. It is a rugged helicopter able to absorb considerable battle damage, for example the main rotor can withstand a hit from a 23mm cannon shell. In the event of a crash, the helicopter's cabin can resist loads of up to 20 G and a 2500ft per minute (28mph) impact is reckoned to be survivable.

The UH-60A carries a crew of three and 11 fully-equipped infantrymen. It is powered by two General Electric T700-GE-700 turboshafts which deliver 1560shp each. Maximum cruising speed at sea level is

160 knots and endurance is 2-3 hours. Maximum take-off weight is 20,250lbs, fuselage length is 51ft and rotor diameter 53ft. An External Stores Support System (ESSS) is being developed, primarily to allow the UH-60A to carry two 450 gallon and two 270 gallon external fuel tanks for ferrying. This will give the helicopter a transatlantic range, staging through the Azores. Alternatively 16 Hellfire antitank missiles or mine dispensers can be carried on the ESSS.

Several NATO armies operate UH-1 troop transports including those of West Germany, Italy, Spain and Turkey. However, the UK and France fly the jointly produced Aérospatiale/Westland Puma. The British examples are operated by the RAF with one squadron in the UK and a second in Germany. France's Pumas are flown by five Régiments d'Hélicoptères de Combat of the Army's Aviation Légère de l'Armée de Terre (ALAT), alongside antitank and reconnaissance Gazelles and Alouette II and IIIs. The Puma is flown by two pilots and can carry up to 20 troops. It is powered by two Turboméca Turmo IVB turboshafts of 1400shp each, giving a maximum speed of 170mph. Range is 365 miles and service ceiling is 15,000ft. Maximum weight is 14,770lbs and dimensions include a fuselage length of 46ft 2in and a rotor diameter of 49ft 3in.

The standard Soviet assault helicopter is the Mil Mi-8 code-named Hip by NATO. Helicopter operations in support of ground forces have increased in

The Soviet Mi-6 Hook heavy transport helicopter (top) can carry small armored vehicles in its cabin. It is here fitted with auxiliary wings to offload the rotor in forward flight. The Bell UH-1 'Huey' (above) was the workhorse troop transport helicopter of the Vietnam War and it remains one of the most important aircraft in service with the US Army. This picture shows a helicopter crew chief in action with the door-mounted M60 machine gun.

A US Army CH-47 Chinook medium lift helicopter (below) supplies a mountain-top position. The British Army Air Corps Lynx AH Mk 1 (bottom left) can be armed with Tow missiles in the anti-tank role, but here carries unguided air-to-ground rockets. The West German Heeresflieger is eqipped with HOT-armed Bö 105 anti-tank helicopters (bottom).

importance as a result of Soviet combat experience in Afghanistan and new units have been formed and existing ones strengthened. The Hip can carry 28 fully-equipped troops or up to 8800lbs of cargo in its cabin, which has rear-loading clamshell doors enabling loads such as small vehicles or antitank guns to be carried. The Mi-8 is armed for laying down suppressive fire during a helicopter assault; the Hip-C carries 128 57mm rockets in four packs mounted on outriggers, while the Hip-E increases this to 192 rockets in six packs, plus four AT-2 Swatter antitank missiles and carries a flexibly-mounted 12.7mm machine gun in the nose.

The Mi-8 is powered by two 1700shp Isotov TV-2 turboshafts (replaced by the 1900shp TV-3 in later aircraft), which gives it a maximum speed of 145mph and a range of 260 miles. Maximum takeoff weight is 26,500lbs and main rotor diameter is 69ft 10in, with a fuselage length of 60ft 1in.

The Hip is supplemented by the Mi-24 Hind-A, which is primarily an assault transport helicopter with a heavy ground attack and anti-armor armament. It has been developed into the more specialized Hind-D and E gunship helicopters. However, in its assault transport form it is a formidable enough machine, which can carry a squad of eight fully-armed troops, in addition to 128 unguided 57mm rockets, four AT-2 Swatters and a nose-mounted 12.7mm machine gun. Hind-A is powered by two 1700shp TV-2 turboshafts (later aircraft have the 1900shp TV-3) and has a maximum speed of around 200mph. Loaded weight is around 22,000lbs and both rotor diameter and the length of the fuselage is 56ft. Hind-A carries a crew of four, comprising two pilots, a gunner in the glazed nose and a loadmaster in the cabin.

Soviet helicopter assault forces will be used to capture key points on the battlefield such as bridges or other bottlenecks. They will probably be used in conjunction with close support aircraft and artillery, or possibly in the aftermath of a tactical nuclear strike. The troops airlifted will be drawn from the motorized infantry divisions, as the Soviet army has no equivalent to the US Army's air cavalry units. Hind-As will probably be used to spearhead such a heliborne assault, suppressing ground fire and dropping troops to secure a landing zone for the following Mi-8s and Mi-6 Hooks carrying heavier equipment.

Most troop transport helicopters also undertake the casualty evacuation role. The US Army assigns special medical evacuation units to this task. They are equipped with the UH-1H, which can carry six litters plus a medical attendant. Rapid evacuation of casualties from the battlefield has considerably reduced the number of fatalities among seriously wounded troops. In Vietnam over 370,000 casualties were evacuated by helicopter and the death rate of those reaching US hospitals was 2.6 percent in contrast to the World War II figure of 4.5 percent. The UH-60A has the same capacity as the

Soviet Frontal Aviation's helicopter armory includes the Mi-8 Hip (right) and Mi-24 Hind-A (bottom) assault helicopters and the Hind-D gunship (below).

UH-1H in the medical evacuation role, while the Soviet Mi-8 can carry up to 12 casualty litters.

Air mobility does not consist simply of airlifting troops, for they must be accompanied by supporting artillery and other heavy equipment, including perhaps armored fighting vehicles. Troops in the field must also be supplied with ammunition and fuel if they are to continue fighting efficiently and often their demands for resupply are so urgent that helicopters must be used rather than slower land vehicles. For these roles heavy and medium lift helicopters are needed to supplement the smaller troop-carrying helicopters.

The US Army's standard medium helicopter is the Boeing Vertol CH-47 Chinook, some 450 of which are in service. These can be used in the troop transport and casualty evacuation roles, carrying 44 troops or 24 casualty litters. However, they are more usually employed as cargo carriers. The CH-47C version can lift a maximum payload of 19,100lbs and can carry a 15,000lb load over a radius of 30 nautical miles. An extensive rebuilding program is underway, which is intended to extend the service life of the CH-47 fleet and to improve the helicopter's performance, payload and maintainability. A total of 436 CH-47A, B and C model Chinooks are to be remanufactured as CH-47Ds, giving them an increased airframe life equivalent to that of a new aircraft. This force is to be increased by a further 91 CH-47Ds, which are to be newly built.

The first rebuilt CH-47D was delivered to the US Army in May 1982. The D variant is powered by two Avco Lycoming T55-L-712 turboshafts of 4500shp each, driving twin rotors. The 30ft long cabin is the same for all Chinook variants, but the maximum externally slung load has increased in weight from the C's 22,700lbs up to 28,000lbs. Maximum speed is 160 knots at sea level and a 14,000lbs internal load can be carried over a 100 nautical mile radius. Loaded weight is 53,500lbs and dimensions include a fuselage length of 51ft and a rotor diameter of 60ft. The Chinook is operated by a crew of three – two pilots and a load-master.

Because the Chinook will usually run out of cabin space before its maximum load is reached, it generally carries its cargo slung beneath the fuselage. This idea has been further developed in the design of the CH-54 Tarhe helicopter, which is in effect a flying crane. Some 80 of these heavy-lift helicopters are in US Army service, performing such specialized duties as positioning heavy artillery, recovering shot down helicopters and aircraft and lifting armored vehicles and engineers' bulldozers and graders. During the Vietnam War these machines recovered some 380 crashed aircraft. The CH-54 has a forward cabin housing three crewmen, including an aft-facing crew member who controls the helicopter when underslung loads are being picked up or landed. Aft of the cabin a boom runs back to the tail rotor, with the engines mounted on top of this structure. There is no main cabin as such, but a detachable pod can be fitted to house passengers or equipment.

The Tarhe is powered by two 4500shp JFTD-12A-4A turboshafts which drive a 72ft diameter main rotor. The fuselage length is 70ft 3in and overall height is 25ft 5in, giving a ground clearance of 9ft 4in beneath the boom. Maximum weight is 47,000lbs, giving a maximum lift capacity of 25,000lbs although 20,000lbs is a more usual load. Speed is 126mph and range is 230 miles.

Among the United States' NATO Allies several countries operate Chinooks, including Canada, Greece, Italy, Spain and the United Kingdom. The RAF's 33 Chinook HC Mk 1s are primarily intended to provide logistical support for the Army, with secondary roles as troops transports and for casualty evacuation. They may also support dispersed site operations by Harrier aircraft. France's ALAT has no helicopter heavier than the Puma, but the West German Army's Heeresflieger has a force of 110 Sikorsky CH-53Gs. These are formed into medium helicopter regiments with one attached to each of the three army corps. Each regiment can move a

lightly-equipped battalion in a single lift.

The Soviet Union's Mil Mi-6 Hook heavy lift helicopter was the largest helicopter in the world at the time of its entry into service at the end of the 1950s and perhaps as many as 400 remain in use. It has a maximum takeoff weight of 93,700lbs with a payload of 26,000lbs. Dimensions include a fuselage length of 108ft 10in and main rotor diameter of 114ft 10in. The Mi-6 is fitted with stub wings to offload the rotor and by using a rolling takeoff rather than a vertical liftoff it can increase its maximum payload. Power is provided by two 5500shp Soloviev D-25V turboshafts, giving a maximum speed of 186mph and a range of over 600 miles with an 8800lbs payload. In the troop transport role 65 men can be carried. However artillery pieces or armored vehicles are more usually carried and Mi-6s were used in this way during the fighting between Ethiopia and Somalia in the Ogaden in 1978. A flying crane version of the Mi-6, the Mi-10 Harke, can lift even heavier loads, although fewer than 100 are in service with Frontal Aviation.

The agile OH-6A scout helicopter (above) serves with US Army units based in the United States and Far East. The CH-54A Tarhe flying crane (left) is invaluable for recovering crashed helicopters and supporting engineer and artillery units. The West German CH-53G medium-lift helicopters (below) can each carry 38 fully-equipped troops.

The Mi-6's successor in the mid-1980s is likely to be the Mi-26 Halo, a 110,000lbs machine with about twice the payload of the Hook. It is powered by two 11,400shp Soloviev D-136 turboshafts, driving a large eight-bladed main rotor. The Mi-26 like its predecessor will operate in concert with troop carrying machines during a heliborne assault, lifting armored vehicles, artillery and perhaps SAMs into the landing zone.

At the other end of the scale from the massive heavy-lift helicopters are the small rotorcraft used for battlefield reconnaissance. These need to be small and agile so that they can make maximum use of terrain masking and nap of the earth flying techniques for cover. In future many will be fitted with mast-mounted sights carried above the rotor, so that they can remain behind cover while observing. Scout helicopters work in conjunction with antitank helicopters, identifying targets and directing attacks. They observe the effects of shellfire for artillery units and scout for armored and infantry formations. They are also useful as liaison machines and as airborne command posts.

The US Army currently operates two types of observation helicopter, the Hughes OH-6A Cayuse (1000 in service) and the Bell OH-58 Kiowa (2000 in service). The OH-6A, which does not at present serve in Europe, is the smaller of the two. It has a maximum loaded weight of 2700lbs, a fuselage length of 23ft and rotor diameter of 26ft 4in. Power is provided by a 317shp Allison T63-A-5A turboshaft, giving a speed of 150mph and a range of 380 miles. Crew comprises a pilot and observer, with accommodation for two passengers. An armament of one 7.62mm XM-27 machine gun with a rate of fire of 4000 rounds per minute is carried.

The larger OH-58A Kiowa, which serves with the US Army in Europe, has a maximum loaded weight of 3000lbs, a fuselage length of 32ft 7in and rotor diameter of 35ft 4in. It is powered by a 317shp Allison T63-A-700, giving a speed of 138mph and range of 300 miles. In addition to its crew of two, three passengers can be carried. Armament is optional, but can include two XM-27 machine guns. The OH-58C version has the more powerful (420shp) Allison 250-C20B engine and is fitted with 'non-glint' windows and infra-red suppressors on the engine exhaust.

The US Army plans to modify 720 OH-58A Kiowas under the army helicopter improvement program (AHIP) to produce a scout capable of working with the AH-64 Apache advanced attack helicopter. The OH-58 AHIP will carry a mast-mounted sight above the rotor, fitted with a TV camera for daylight operation, imaging IR for night observation and a laser rangefinder and designator. Using the laser, the OH-58 AHIP will be able to mark targets for the AH-64's Hellfire antitank missiles, or for Copperhead laser-guided artillery shells. Other improvements to the OH-58A include fitting the more powerful Allison 250-C30R, increasing payload by 550lbs and an armament of Stinger anti-

aircraft missiles to provide prototection against attack helicopters and CAS aircraft.

The most important scout helicopters in service with the European NATO armies are the Anglo-French Gazelle and West Germany's Bö 105. The Gazelle serves with ALAT in both the reconnaissance and anti-tank roles (armed with HOT antitank missiles). The UK's Army Air Corps uses the Gazelle for observation and liaison. Powered by a 590shp Turboméca Astazou turboshaft, the Gazelle has a speed of 190mph and a range of 415 miles. Maximum loaded weight is 3970lbs and dimensions include a fuselage length of 31ft 3in and a rotor diameter of 34ft 5in.

West Germany's Heeresflieger uses the Bö 105 for observation and antitank duties and it also serves in the Netherlands. It is powered by 400shp Allison 250-C20 turboshafts and has a maximum speed of 165mph and 400 mile range. With a maximum weight of 5070lbs, the Bö 105 has a fuselage length of 28ft and a rotor diameter of 32ft 3in. It can carry four troops, plus the pilot, and in the antitank role it carries six HOT missiles.

Surprisingly, the Soviet Union has no equvalent to NATO's large forces of scout helicopters. This deficiency is as much a result of differing operational philosophies, as of the apparent inability of Soviet designers to produce a small and agile scouting helicopter. Soviet helicopter tactics emphasize the use of heavily armed assault and gunship helicopters in forays behind the enemy lines, intended to cause disruption and to seize key points. NATO by contrast intends to counter a blitzkrieg-type tank assault by various carefully planned anti-armor measures which require the accurate intelligence that scouting helicopters can provide. Therefore the nearest Soviet equivalent to the Western scout helicopters is a force of elderly Mil Mi-1s, Mi-2s and Mi-4s used for miscellaneous duties such as staff transport and liaison.

NATO's reliance on timely intelligence of enemy moves is reflected in the various electronic reconnaissance helicopters and fixed wing aircraft in service with the US Army or under development. The Gruman OV-1 Mohawk is a twin-turboprop fixed wing reconnaissance aircraft which provides battlefield reconnaissance at division level. Operated by a crew of two, the Mohawk can fly from short and roughly-surfaced airstrips close to the battlefront. It carries various sensors, including SLAR, cameras and infra-red. Some Mohawks have been modified for electronic reconnaissance. Powered by two 1400shp Lycoming T53-L-701 turboprops, the Mohawk has a top speed of 290mph and a range of over 900 miles. Maximum loaded weight is 18,100lbs and dimensions include a span of 48ft and a length of 41ft.

An altogether more specialized Army fixed-wing aircraft is the Beech RU-21 Ute, which is used for signals intelligence. By monitoring enemy radio transmissions, information on his order of battle can be gleaned and preparations made to jam his signals

A US Army OV-1 Mohawk fires target-marking rockets during a surveillance mission. OV-1s are normally unarmed.

traffic. Furthermore by studying and analyzing his radio traffic, it can be deduced that certain levels and patterns of radio transmissions will presage various sorts of activity. All of this information can be gathered without breaking enemy ciphers. Should the messages intercepted by deciphered, then the intelligence haul will be so much the greater.

The RU-21 Ute is derived from the Beechcraft King Air executive transport. It is powered by two UAC 500shp PT6A-20 turboprops, giving it a speed of 250mph and a range of 1160 miles. Maximum weight is 9650lbs, span is 45ft 10in and length 35ft 6in. The crew comprises two pilots and two electronic systems operators. Prominent antennae are fitted to the wings and rear fuselage forming part of the radio intercept equipment.

A version of the UH-60A Black Hawk helicopter is being developed for signals intelligence and jamming. Designated the EH-60A, it carries 1800lbs of mission equipment and is fitted with a large whip aerial beneath the fuselage and four diplole aerials aerials on the tail boom. The US Army requirement is for 77 EH-60As.

Two altogether more ambitious programs are also based on the UH-60 airframe. The EH-60B Sotas is a stand-off target acquition system, which carries a radar capable of detecting moving targets, such as tanks and other vehicles, deep in enemy territory. The informa-

tion is passed to a ground station by data link. However, because of development problems and unacceptable cost increases, it now seems unlikely that the 61 needed by the US Army will be produced. The other UH-60 development is the Airborne Radar Jamming System, which seeks to fit a simpler and lighter version of the EF-111's ALQ-99 ECM system to the helicopter. This is needed because of the build-up of Soviet radar-directed SAMs and AA guns on the battlefield.

The specialized attack helicopter was the outcome of experience in Vietnam, where it was found necessary to provide heavily-armed escorts for troop-carrying UH-1s and CH-47s. A two seat attack helicopter, the AH-1G Huey Cobra was produced, using the power-plant and transmission system of the UH-1. Armed with a turret mounted 7.62mm GAU-2B minigun (later superseded by the XM-28) and a 40mm grenade launcher, the AH-1G entered combat in 1967 and was quick to prove its worth. It was used defensively to escort other helicopters and lay down suppressive fire during landing operations, carrying up to 76 FFAR unguided 2.75in rockets. It was also used in offensive roles, hunting for enemy troops and providing close air support when the enemy was too close to friendly forces

to allow the use of USAF CAS aircraft. The great advantages of the AH-1G were its agility, its ability to use ground cover and the excellent visibility from its cockpit.

The AH-1G's characteristics were found to be admirably suited to the antitank helicopter role in Europe and approximately 100 AH-1Gs were modified to AH-1Q standard by fitting the BGM-71 TOW antitank missile. These aircraft retained the chin turret, with 7.62mm gun and grenade launcher. However, they were only an interim solution and the US Army has standardized on the improved AH-1S, 987 of which are due in service by 1984. The AH-1S replaces the earlier Huey Cobra's 1250shp Lycoming T53-L-13 power-plant with the more powerful 1825shp T53-L-703, giving sufficient reserves of power to accelerate from the hover to 150 knots in 11 seconds. The nose turret armament is retained and the stub wings can carry eight TOW missiles, plus two 19-rocket FFAR pods.

Basic characteristics of the AH-1S include a maximum takeoff weight of 10,000lbs, a fuselage length of 44ft 5in and a main rotor diameter of 44ft. Cruising speed is 120 knots, rate of climb is 1600ft per minute and range 275 nautical miles. Armament is to be upgraded by replacing the 7.62mm minigun and grenade launcher with either an M197 three barrel 20mm cannon, with a rate of fire of 750 rounds per minute, or the 30mm Hughes Chain Gun (700 rounds per minute).

The TOW antitank missile has a maximum range of 12,300ft and is effective down to 200ft. The missiles are aimed by the gunner using a chin-mounted sighting turret giving a $\times 2$ or $\times 13$ magnification. This optical sight is to be augmented by a FLIR sight to give a limited night fighting capability. The gun turret can be controlled by both pilot and gunner using helmet mounted sights and a fire control computer, linked to a laser tracker, provides weapons release instructions for rockets and gunfire impact points projected onto a HUD.

Operating over the European battlefield the AH-1S will be exposed to numerous threats, including ZSU-23-4 radar-controlled AA fire, SA-7 man-portable, IR guided SAMs and the more powerful, mobile SA-4s and SA-6s. The use of terrain-masking to provide cover is an effective tactic to counter these weapons, but the helicopter must expose itself while firing and guiding its TOW missiles. A radar warning receiver will warn of the presence of guns and radar-guided missiles, allowing the crew to take evasive action. Heat-seeking IR missiles are countered by a 'black hole' suppressor fitted to the engine exhaust, which reduces the IR signature, and by an IR jamming beacon which will confuse the missiles' homing heads. The crew seats are surrounded by protective armor and the narrow, head-on cross-section of the AH-1 (fuselage width is 38in) will also help.

AH-1 battle tactics depend on nap-of-the-earth flying and use of ground cover. The terrain of Central and Southern Germany where they will operate is especially suited to such tactics. Because of dense AA defenses surrounding Soviet armored formations, the AH-1s will seek to engage from long range and they will co-ordinate their tactics with friendly artillery and A-10 aircraft. As the territory that they will fight over is the same as that which they use for peacetime training, AH-1 crews will have the considerable advantage of good local knowledge of the terrain. Normally they will operate in groups of five, directed by three OH-58 scouts. Short-range, high-powered radio transmissions are difficult to jam, but AH-1 crews keep the use of radio to a minimum by careful preflight briefing and the use of hand signals.

The AH-1S has a limited ability to fly and to fight at night. The new FLIR-Augmented Cobra TOW sight (FACTS) will give the gunner a night-firing capability,

but the pilot needs to wear image-intensifying night goggles. As these need to be refocused if he wants to look away from his instruments and out of the cockpit, the system is rather inflexible. A much better night fighting capability will be provided by the Hughes AH-64A Apache, which is due to become operational in 1985.

The AH-64A offers a great deal more than the ability to fly nap-of-the-earth sorties at night or in bad weather, however. It will be armed with up to 16 Rockwell Hellfire laser-guided missiles giving it a 'fire-and-forget' capability, unlike the TOW which has to be guided to its target from the launch helicopter. Used in this way, the AH-64A will remain behind cover while its target is designated by laser from the ground or from a scout helicopter. The Apache can then launch its Hellfires from cover, lobbing them over the intervening terrain to the designated target.

The Apache is powered by two General Electric T700-GE-701 turboshafts rated at 1690shp each. Maximum speed is over 150 knots, rate of climb is 1300ft per minute and endurance is over two hours. With a maximum weight of 14,000lbs, the AH-64A has an overall length of 56ft 9in and a main rotor diameter of 48ft 8in. Apart from the Hellfire missiles, armament options include a 30mm Hughes Chain Gun with 1200 rounds of ammunition and up to 76 unguided 2.75in FFAR rockets.

This early AH-1G Hueycobra (left) carries gun and rocket pods in addition to its turret-mounted armament. The AH-64 Apache's 30mm Chain Gun is mounted on a trainable turret beneath the fuselage (below).

A YAH-64 Apache prototype fires a salvo of 2.75in FFAR unguided rockets.

The Apache's night fighting ability is due to the Pilot's Night Vision Sensor (PNVS) and the gunner's Target Acquisition Designation Sight (TADS). The PNVS comprises a FLIR which presents an IR image of the terrain in front and to the side of the helicopter. This is displayed on a monocle sight, which the pilot can position in front of either eye, and flight instrument data is also projected onto the sight. The PNVS makes possible nap-of-the-earth flying at night and in poor visibility. It is complemented by TADS, which gives the gunner a day and night target sight, plus a laser tracker, target designator and rangefinder.

Apache has numerous built-in survivability features,.which should help to reduce its vulnerability. It has a low silhouette in profile and a small head-on cross section. The main and tail rotors are quieter than those of most helicopters and the engine's IR signature is reduced by a 'black hole' suppressor. Crew seats are armored, as are key engine components and fuel cells. Each of the four rotor blades can withstand a hit by a 23mm cannon shell and the main gear box will continue to function for up to an hour after losing its oil.

These sophisticated capabilities do not come cheaply. The US Army plans to acquire a total of 446 Apaches at a cost of $9.5 million apiece. However, Congressional approval has only been obtained for an initial production batch of 11 and there is some doubt whether all 446 will be procured. In service Apaches will supplement rather than replace the AH-1S force.

The antitank helicopter has been developed independently of the United States by three major European NATO Allies. The British Army Air Corps is basically organized into two-squadron regiments, with one squadron flying 12 Gazelle scouts and the other (when re-equipment is completed in the mid-1980s) with 12 TOW-armed Westland Lynx AH Mk 1s. The five AAC regiments serving with British Army of the Rhine in Germany have all re-equipped, while UK-based units continue to fly Westland Scouts armed with the Nord SS-11 missile.

West Germany's Heeresflieger plans to equip three antitank regiments with the PAH-1 (Bö 105P) helicopter, 212 of which are on order. Each regiment will have a strength of 56 helicopters, with a regiment assigned to each of the three army corps. The PAH-1 is

armed with six Euromissile HOT 13,000ft range anti-tank missiles and German crews have achieved a success rate of better than 90 percent during practice firings. The basic tactical unit of PAH-1s is the seven-aircraft Schwarm and a daily sortie rate of up to five per aircraft is anticipated.

France's main antitank helicopter is the HOT-armed Gazelle, which serves alongside the older SS-11-armed Alouette III in the six Régiments d'Hélecoptères de Combat (RHC). Each RHC has three antitank-helicopter escadrilles and some 160 Gazelles are on order to equip them. France and West Germany may co-operate to produce a new antitank helicopter armed with advanced antitank missiles and with a night and adverse weather capability.

The heavily-armed Hind-D and -E variants of the Mi-24 represent the Soviet Union's nearest equivalent to the NATO antitank attack helicopters. They carry the pilot and gunner in individual cockpits in the nose section, with a four-barrel 23mm cannon mounted beneath the nose. Otherwise the armament is similar to that of Hind-A, namely four 32-round 57mm rocket pods and four AT-2 Swatter antitank missiles. The 1.5 mile range Swatters are replaced by the tube-launched AT-6 Spiral on the Hind-E, with range increased to some 5.5 miles. The Hind gunship's weapons aiming sensors are believed to include radar, low light TV and a laser rangefinder. As the gunships retain the troop carrying capabilities of the earlier Hinds, they are far more flexible machines than the NATO attack helicopters. Their combat roles may include antitank warfare, helicopter assault escort and fire support, close air support, helicopter assault and anti-helicopter operations. However, this versatility has incurred a penalty, as the Hind-D with its 22,000lbs normal take-off weight is considerably larger and more cumbersome than any Western attack helicopter.

Because helicopters are so extensively used on the battlefield they are likely to become a prime target for their own kind and for CAS aircraft. The Hind-D/E's 23mm cannon with its 3200rpm rate of fire is considered a potent anti-helicopter weapon. The Apache's Hughes 30mm Chain Gun falls into the same category. Whereas the US Army has found such air-to-air missiles as the Sidewinder to be unsuitable for helicopter use, the lightweight Stinger IR-guided SAM has proved to be compatible with the attack helicopter and AH-64As may carry Stingers. Such armament experiments could eventually lead to a specialized anti-helicopter helicopter.

Anti-helicopter tactics could combine the use of attack helicopters and CAS aircraft. In the United States trials have been flown combining A-10s and AH-1s in anti-helicopter teams with some success. The Luftwaffe's Alpha Jet CAS aircraft have an anti-helicopter role, intended to counter heliborne assaults by Mi-8s and Mi-24s and also to shield NATO antitank helicopters from attack by Hind-D/E gunships.

The Balance of Forces

Who will win World War III in the air? Numerous factors will affect the way in which an air force performs in action and by no means all of them are quantifiable or subject to assessment by clinical analysis. Nevertheless, such considerations as the quantity and quality of equipment, peacetime training standards, motivation of personnel and readiness to react to the often unexpected demands of war are all indications of an air force's worth. This chapter attempts to assess the capabilities of the NATO and Warsaw Pact air forces and to indicate their strengths and weaknesses.

The race may not always go to the swiftest, nor the fight to the strong – but as a cynic once perceptively remarked, that is the way that you bet. Direct comparisons between the numbers of aircraft of the Warsaw Pact and NATO air arms can be misleading. It is obvious that such comparisons ignore the quality of the machines, but many other factors must also be considered. For example the serviceability of aircraft and the standards of maintenance must be assessed, because an air force that is strong on paper may have a high proportion of its aircraft grounded for servicing. Similarly an apparently well-equipped air arm may not have reserves of equipment and trained manpower to fight a sustained campaign. Furthermore a tally of numerical strength cannot show how well the aircraft are deployed to undertake this wartime role. They may have to fly thousands of miles to the theater of war.

Therefore a comparison of strengths is by itself a crude and often misleading indication of the outcome of a conflict, yet it must be the starting point of any comparative analysis. In bare statistics the Soviet Union's strategic forces comprise some 7000 nuclear warheads, which range in yield from 200 kilotons up to 24 megatons. They are delivered by 1398 land based ICBMs, 950 SLBMs aboard 70 modern submarines and some 150 long range bombers. The bomber force can be doubled if the Tu-26 Backfire is regarded as an inter-continental-range bomber. The comparable figures for the US strategic forces are 1052 land-based ICBMs, 520 SLBMs aboard 32 submarines and 376 manned bombers. The introduction into service of the air-launched cruise missile has added a further element to the equation, with the United States planning to procure perhaps as many as 4300 ALCMs by 1991.

Although a comparison of delivery vehicles would imply that the Soviet Union has the edge in strategic weapons, this is not so. Because most of the United States' missiles carry multiple, independently-targetable warheads, the US missile force alone can deliver more than 7000 warheads, with the manned bomber force contributing several thousand more. The fear that the US land-based strategic arsenal can be wiped out in a massive Soviet first strike, must be counterbalanced

A pair of US Navy F-4 Phantoms shadow a Soviet Tu-16 Badger as it overflies the carrier USS Kitty Hawk.

by the confidence that submarine-based missiles will survive as a secure second-strike force. Furthermore, the arguments put forward in favor of the closely-spaced basing mode for the new MX ICBM – that the nuclear explosions from the first warheads to detonate will knock following warheads off course – also tend to suggest that an effective first strike against the existing Minuteman force would be very difficult. It appears that the Soviet Union will use its submarine-launched missiles as part of its first strike force, targeting them on SAC airfields and headquarters. If this is so then the Soviet Union, unlike the United States, has no secure second strike capability, but must rely on the possibly ineffective expedient of reloading cold-launch silos.

Both the Soviet Union and the United States are well aware that in an all-out nuclear war there will be no victors. Strategic forces primarily undertake a deterrent role and if one side suggests its willingness to fight a nuclear war (as the Soviet Union has often appeared to do), this is more likely to be an attempt to strengthen its deterrent posture, than a reflection of its true doctrine. Nuclear forces are a powerful political lever which both of the Superpowers make use of in their dealings with adversaries and allies alike. Paradoxically these immensely powerful weapons exert most influence when they are not used – indeed for most military requirements they are totally unusable.

If this assessment is correct then it is at the tactical level that the outcome of World War 3 will be decided. Tactical nuclear weapons, apart from such specialized applications as nuclear depth charges, are simply deterrent forces at a theater level. A commander would be most reluctant to employ them, unless the situation had deteriorated to such an extent that their use was the only way of averting a catastrophic defeat. Modern conventional weaponry is so powerful that in many instances tactical nuclear weapons would not be very much greater in destructive effect. Provided that one side does not perceive that it can use tactical nuclear weapons with impunity, or that the opponent is not facing irrevocable defeat, then tactical nuclear weapons will play primarily a deterrent role. The battle will be fought with conventional weapons to a mutually accepted stalemate – or to the point when nuclear weapons are introduced and thereafter its course will be incalculable. There is no experience of nuclear war to draw on and it may well be that a tactical exchange will swiftly escalate into all-out war.

The possession of nuclear weapons is therefore no substitute for well-equipped and highly trained conventional forces. Non-nuclear forces have the great advantage that they are usable at any level of conflict from counter-insurgency up to a major war. Both the United States and the Soviet Union have strong and effective tactical air arms. The Soviet air force's Frontal Aviation comprises some 4800 fixed wing combat aircraft, supported by 250 transport aircraft and some 3500 helicopters. The comparable figure for the USAF is just over 3000 tactical aircraft, 2800 of which are tactical fighters. However, to this total must be added the tactical aircraft embarked aboard US Navy aircraft carriers (approximately 1000) and the 400-odd combat aircraft of the US Marine Corps. These figures give the Soviet Union a small margin of superiority in tactical aircraft. Yet if the United States' NATO allies with 3000 tactical aircraft and the non-Soviet Warsaw Pact's air forces with 2400 tactical aircraft are included in the reckoning then it can be seen that the two sides are roughly equal in strength, both having some 7200 tactical aircraft.

Peacetime deployment of tactical air forces presents some problems for the United States. A proportion of the force is based on forward airfields in Europe and Pacific, but most tactical fighters will have to deploy to the theater of war from bases in the continental United States. All tactical fighters are capable of refueling in flight, but shortages of tanker aircraft or intermediate landing facilities could hamper the rapid reaction of USAF's Tactical Air Command to a crisis. Apart from reinforcing NATO's Central Front in Europe, US tactical fighters may have to deploy to the Middle East or to Japan and South Korea. The Soviet air force has the advantage of operating on interior lines of communication, with numerous tactical airfields available to facilitate the rapid movement of air forces between their peacetime bases and theaters of war.

Because the Soviet Union operates on interior lines of communication, whereas the United States is geographically isolated from most centers of conflict, the latter's need for air transport is therefore immeasurably greater. The bulk of the Soviet air force's 600 medium and long range transport aircraft will be used tactically for logistical support of the armies and to lift the airborne divisions. The USAF's Military Airlift Command with a strength of some 830 transports must divide its resources between tactical and strategic airlift. The rigidly state-controlled civil air operations of the Soviet Union provide a far more efficient military transport reserve than does the United States' Civil Reserve Air Fleet. Aeroflot can provide some 200 Il-76 and An-12 transport aircraft, 1100 medium and long-range passenger transports and several thousand short-range transports and helicopters to bolster the military transport forces. The American CRAF in contrast will make available only 109 long-range cargo aircraft, 215 long-range passenger transports and 84 shorter-range transports.

Naval aviation is an area of great strength for the NATO powers as their forces and capabilities outmatch those of the Soviet Union by a wide margin. However, while control of the sea is essential for the United States and her allies if they are to prosecute a successful war, for a land power like the Soviet Union it is of lesser importance. This is not to deny that US naval power – and in particular the strike carriers and SLBM-carrying submarines – poses a very worrying threat to the Soviet

The F-16XL, which has been modified as a tail-less delta, powered by a 25,000lb-thrust F101 turbofan, may be the USAF's next interdiction aircraft.

Union. In terms of numbers, the difference between the US Navy's 1800 combat aircraft and the Soviet Union's 1400 naval aircraft is relatively small. The essential difference lies in the United States' fleet of strike carriers, which enable over half the service's strength to be carried aboard mobile airfields to any part of the world's oceans where they are needed. In contrast the greater part of Soviet naval air strength is tied to its land airfields.

Air defense forces are best assessed in relation to the threats they are intended to counter. The USAF's force of 300 interceptors is dwarfed by the Soviet air defense arsenal of 2500 aircraft and 10,000 SAMs. Yet when the USAF's 300 interceptors are set against the Soviet Union's 150 long-range bombers, the force looks less ridiculous. Its strength is probably about right, but the interceptors themselves are for the most part long over-due for replacement (the same can be said for the Soviet strategic bomber aircraft, though). The balance be-tween 2500 Soviet interceptors and 376 US manned bombers looks less realistic, until it is remembered that the interceptors will also have to deal with far-ranging tactical aircraft and, with the introduction of the cruise missile, a twenty-fold increase in the threat from each B-52 so equipped.

The US has a considerable advantage in battlefield transport and attack helicopters, with a strength of some 9000 rotary-winged aircraft, compared with the Soviet Union's 3500 machines. The advantage is not simply in numbers, though, for the Soviet helicopters are for the most part large and cumbersome – albeit heavily armed in many cases. They are likely to prove far more vulnerable than their lighter and more agile Western opponents. However, following the criteria used to assess air defense forces, the antitank element of the US force (around 1000 AH-1s equipped with TOW) should be set against the Soviet Union's 50,000-strong tank force.

In general Soviet technology lags behind the West, especially in such areas as microelectronics and com-puters, so important in the design of combat aircraft and missiles. When Western analysts examined the MiG-25 Foxbat which the defecting Soviet pilot Viktor Belenko flew into Hakodate, Japan, in 1976, they found that much of the structure was of steel rather than titanium, with none of the carbonfiber composites to be found in contemporary Western aircraft. The avionics were similarly dated, making use of vacuum tubes rather than solid-state printed circuits. However, the aeroplane performed the task which it was designed to undertake and its performance was such as to cause considerable concern in the West.

While Soviet aircraft and their equipment may appear crude by Western standards, they nonetheless perform effectively. They often cannot match their opponents in technological sophistication, for in spite of the widely publicized advances in weapons tech-nology made by the Soviet Union during the last decade, it still lags behind the United States in most areas. However, Western sophistication – some critics would say needless complexity – is often offset by sheer numbers. The Soviet Union has consistently produced fighter aircraft at the rate of some 1300 per annum for the past five years. Since the early 1970s it has replaced limited-capability MiG-21 tactical fighters with the truly multi-role MiG-23/27 Flogger. While this aircraft does not carry a radar to match those of the F-15 or F-16 and is not nearly so agile as these American fighters, it is available in sufficient numbers to swamp the opposition. Tactical evaluations in the United States have shown that in dogfights involving six or more aircraft, advanced fighters lose much of the advantage from their superior weaponry, propulsion and aerodynamics and the fight degenerates into a slugging match.

Instructive comparisons can be made between the performance characteristics of the major tactical fighter aircraft flown by the USAF and the Warsaw Pact air forces:

Aircraft	MiG-21bis	MiG-23	F-16A	F-15C
Warload	3300lb	3300lb	14,00lb	16,000lb
Max Speed	Mach 2	Mach 2+	Mach 2+	Mach 2.5+
Ceiling	48,000ft	58,000ft	58,000ft	65,000ft
Runway requirement	2800ft	3000ft	2500ft	2500ft
Tactical Radius	560 miles	800 miles	575 miles	750 miles
Armament	one 23mm cannon four AAM	one 23mm cannon six AAM	one 20mm cannon four AAM	one 20mm cannon eight AAM

What is immediately apparent is the higher warload of the USAF fighters, which gives them a very valuable dual air-to-air and air-to-ground capability. The comparisons of tactical radii can be misleading, as the American fighters are able to extend their range by inflight refueling, whereas no Soviet tactical fighter has yet been equipped for this. The slightly better runway-length requirements of the American fighters must be offset against the Soviet types' rough field capability and the greater availability of tactical airfields in the Warsaw Pact nations. Against the Soviet fighters' heavier caliber cannon armament, must be set the respective rates of fire. The Soviet GSh-23 fires at rather more than 3000rpm, while the US M61 vulcan cannon's rate is 6000rpm. Missile armament is broadly comparable; the MiG-23 and F-15 both carry a mix of medium and short range missiles. However, the American AAMs perform better and have greater ranges. What aircraft performance charts often fail to reveal is comparative maneuverability. The MiG-21bis has a rate of climb closely comparable with that of the F-16, but the American fighter can turn twice as fast as the MiG-21 at sea level. Perhaps the most significant factor in comparing these Soviet and US aircraft is numbers in service. The Soviet air force's Frontal Aviation has around 1400 Floggers (including MiG-23s and MiG-27 ground attack fighters) and 1300 Mig-21s in service. The USAF's tactical fighter wings have around 300 F-16s and 400 F-15s.

There are many ways in which the US lead in electronic and computer technology can be used to offset a numerical disadvantage. Advanced ground-based and airborne radars can ensure that fighter assets are deployed in the most economical and effective way.

Similarly such capable stand-off reconnaissance systems as the Lockheed TR-1A can warn of an impending armored assault, indicate its strength and identify its direction. Such intelligence will be of priceless value in planning defensive tactics against a vastly stronger tank force. NATO has always been forced into the position of countering superior Warsaw Pact numbers with fewer but more advanced weapons systems. This is why Western defense planners find the recent Soviet improvements in weaponry so worrying, for they fear that the Soviet Union is closing the technological gap while maintaining its numerical advantage. This is obviously a danger that has to be guarded against, but in the air at any rate the Western technological lead is apparent, even if this position can no longer be regarded as unassailable.

Sophisticated equipment obviously is of little use unless it is properly maintained. The USAF has been criticized for the low standards of its aircraft maintenance, and the consequent high unserviceability rates of its warplanes. This concern, while not unfounded, is probably exaggerated. A good deal of field maintenance is relatively undemanding, involving the check-out of aircraft systems, often using built-in test equipment, and the replacement of faulty parts by line-replaceable units rather than repairing them on the aircraft. By their nature modern combat aircraft are very complex assemblies of advanced equipment and so a certain degree of unserviceability is unavoidable. Soviet aircraft have the reputation of being easily serviced by relatively unskilled personnel, a very necessary characteristic of the equipment of a largely conscript-manned air force.

One area in which the NATO air forces are seeking to redress the massive imbalance between Western and Warsaw Pact ground forces is in all-weather attack. At present the interdiction campaign will continue in all weathers and throughout the hours of darkness. However, close air support and helicopter antitank operations are now largely daylight and fair weather activities, whereas ground combat is likely to continue for 24 hours a day. By developing pod-mounted FLIR, radar and laser designator/seekers, the current force of USAF CAS aircraft can be rapidly converted to all-weather operation. The next generation of attack helicopters will also be equipped to fight in darkness and limited visibility.

By day or night tactical aircraft will be faced by concentrated Soviet ground defenses, including radar directed AA guns and SAMs. The USAF has developed a unique defense suppression capability based on the F-4G Wild Weasel aircraft armed with anti-radar missiles. Operating in concert with the EF-111A tactical jamming aircraft, the Wild Weasels can considerably degrade the performance of enemy ground-based air defences and open the way for the attack aircraft. Soviet EW capabilities are also considerable, but in an intense ECM environment (with all communications jammed)

The Tu-26 Backfire bomber poses a serious threat to NATO's land and sea supply lines.

the individualistic Western combat pilot is likely to perform better than his rigidly disciplined Soviet counterpart. The dependence of the Soviet serviceman on orders from a higher authority is a weakness that the NATO forces can exploit to considerable advantage.

There is much then that technology can contribute to redressing the balance of conventional forces in Europe, which is at present so heavily weighted in favor of the Soviet Union. One of the great problems for NATO is that of preacetime deployment and the related question of readiness. The Soviet air armies are for the most part concentrated in the westernmost military districts of the USSR and in East Germany, Poland, Czechoslovakia and Hungary. In contrast many USAF tactical air units are based in the United States. This creates a potentially dangerous imbalance of forces in the European theater, most especially on NATO's Central Front, which is exacerbated by the overwhelming Soviet superiority in ground forces. However, USAFE can be rapidly reinforced in time of crisis, provided that there is the political will to order a timely reinforcement.

The peacetime readiness states of tactical air units is generally of a high order and reinforcement of the NATO central region is exercised frequently, notably during large-scale US Reforger (*return of forces to Germany*) exercises and by frequent smaller-scale detachment of USAF squadrons or wings to Europe. Such deployments not only provide practice of reinforcement procedures, but they also help to accustom US-based aircrews to the very different operating environment of Western Europe. Reserve forces play an

important part in bringing the US air forces up to a war footing. The Air National Guard and Air Force Reserve contribute 34 percent of the USAF tactical fighter force, 57 percent of reconnaissance assets and 60 percent of tactical airlift. Similarly reserves bolster the US Navy's carrier air wing strength by 14 percent and Marine Corps light attack squadrons by 34 percent. In general the reserve forces are well trained and well-equipped, often flying the same aircraft types as the regular units.

An important increment of strength to US and Soviet forces comes from the NATO and Warsaw Pact allies – and possibly too from friendly forces outside these alliances. Within NATO are to be found for the most part well-equipped and highly-trained air forces. Those of France, West Germany and the UK are especially effective. Although France remains a member of the alliance, no French forces are assigned to NATO commands, but there is little doubt that France would fight alongside the other members in the event of war. Less well-equipped NATO air forces are those of Portugal and Turkey, while the continuing rift between Turkey and Greece further weakens the NATO Southern Front. Economic recession has affected the modernization plans of most NATO air forces and Belgium has even been forced to restrict her pilots' flying to a level well below the NATO minimum as an economy measure.

While the quality of equipment and crew training is

generally very good amongst the NATO allies and the motivation of personnel is high, inter-operability and standardization between the various air arms is often a problem. During the 1970s it was even difficult for an aircraft of one NATO air force to refuel at the air bases of another member, because of differences in the fuel used and in the equipment. Many of these basic problems have been solved. It is now common practice for NATO units to operate from another member-nation's air bases on exchange visits.

However the basic problem of lack of standardization remains. Each NATO member procures its aircraft independently and so there is a wide range of different aircraft in NATO service performing broadly similar roles. Attempts to achieve a greater degree of standardization have met with only limited success. The Starfighter program was perhaps the most successful. Yet of the nine NATO nations which acquired the F-104 Starfighter, only four have elected to follow it with European-manufactured F-16s. Three major NATO members, West Germany, Italy and the United Kingdom, have combined resources to produce the Tornado strike fighter and a number of bilateral agreements between member nations have produced jointly designed and built combat aircraft. Nonetheless, in comparison with the air forces of the Warsaw Pact, NATO displays a conspicuous lack of standardization in equipment and even in tactics. Consequently scarce funds and resources are wasted in duplication of effort.

The Soviet Union's allies in Eastern Europe suffer from no such problems in the selection of equipment or the formulation of tactics. Their air forces are almost

These F-5E Tiger IIs are used by USAF Aggressor squadrons to simulate Soviet fighters during highly-realistic air combat training at Nellis AFB, Nevada.

exclusively armed with Soviet warplanes and their personnel are trained in Soviet tactical doctrines. They contribute some 2400 fixed wing aircraft and 800 helicopters to the Warsaw Pact air forces. For the most part these are rather dated Soviet machines, such types as the MiG-23 only reaching the Warsaw Pact allies slowly and in small numbers. They carry out air defense, air superiority, ground attack and reconnaissance roles. The counter-air roles predominate, as the Soviet Union regards its Warsaw Pact allies as useful buffer zones between Soviet territory and the NATO airfields. Indeed the air defenses of the satellite nations constitute six additional districts within the Soviet air defense network and these are commanded by Soviet officers. There are only small national aircraft industries in Eastern Europe, although Czechoslovakia has designed and built L29 Delfin and L39 Albatros jet trainers for the Pact (except for Poland, which uses its own TS11 Iskra). It is to be doubted whether the Warsaw Pact allies would be enthusiastic participants in a Soviet-inspired conflict. Nevertheless there is no reason to suppose that they would refuse to fight alongside the Soviet air force to which they would most probably be subordinated in wartime.

The air fighting of World War 3 is unlikely to be confined to the air forces of the Warsaw Pact and NATO alliances. A possible participant is the People's Republic of China, an implacable enemy of the Soviet Union since the early 1960s. China is the world's third-ranking air power, having some 4500 aircraft. However, their effectiveness is likely to be poor. The main tactical fighter types are license-built versions of the MiG-17 and MiG-19, both of which have been withdrawn from Soviet service as obsolete. A Chinese attempt to build a copy of the early-model MiG-21F has not been very successful and only small quantities of this aircraft are in service. They reportedly require an engine change after 75-100 flight hours. A more successful venture has been the A-5 Fantan, an adaptation of the MiG-19 with an improved ground-attack capability, several hundred of which are in service. However for all its size, the Air Force of the People's Liberation Army is unlikely to seek a confrontation with the Soviet Union. Its combat performance against Vietnam in 1979 was disappointing and, like the other branches of the Chinese armed forces, it is badly in need of an infusion of modern equipment.

The performance of air forces in combat is not simply a question of the technical proficiency of equipment or the numbers of aircraft in service. However the human element is far more difficult to assess and quantify than hardware. The NATO air forces in the main appear to be manned by highly-motivated individuals with a genuine commitment to the preservation of the democratic way of life. Much of the training received by combat pilots in the West stresses the value of individual initiative at junior levels of command and emphasizes teamwork rather than the rigid subordination to higher authority that can stifle self-motivation. However, it may be that life in a democracy is a poor schooling for the rigors of modern warfare and that the tough airmen trained under the rigid discipline of the Communist states will prove to be very formidable opponents.

The picture presented by the defecting Soviet airman Lieutenant Viktor Belenko of life in the Soviet air forces (he served in the Soviet Air Defense Force) is a bleak one. He presents the Soviet air force as a brutally-disciplined service, grossly inefficient and manned for the most part by discontented conscripts who try to drown their grievances in orgies of drunkenness. There may well be elements of truth in this view. However any defector's testimony, however well-intentioned it may be, must be suspect because it comes from a malcontent. By Belenko's own testimony, the position of a military pilot in Soviet society is a privileged one and, as a result, the average Soviet airman could well be a highly-motivated and loyal Communist. If he is not then it will not be due to any lack of political indoctrination within the Soviet armed forces.

The other aspect of the human equation in air warfare is training. However well equipped an air force may be, its performance will in the final analysis depend upon the skill of the pilots in flying and fighting their aircraft. The United States has developed a system of combat training that is unlikely to be bettered by any other air power. During combat operations in Vietnam it was discovered that a pilot's first ten operational sorties were a crucial testing period. If he could live through these, thereafter his chances of survival increased dramatically, as he had become attuned to the demands of combat. Tactical Air Command's Red Flag exercises aim to give American and allied military pilots this vitally important experience in peacetime.

In an area of the Nevada desert as large as Switzerland, United States fighter squadrons are pitted against realistically simulated Soviet air and ground defenses. The Red Flag exercises typically last for a period of four weeks and can involve 2500 sorties by the participating aircraft. Live ordnance is dropped onto various simulated targets, which range from an industrial complex to a Soviet armored division. The opposition is provided by radars operating on the same frequencies as Soviet AA gun and SAM radars. Aircraft are 'shot down' with video camera recorders. In the air the opposition comes from USAF tactical fighter squadrons and from the Aggressors – especially trained and equipped fighter squadrons which operate in the same manner as Soviet fighter units.

The Aggressor squadrons like the Red Flag exercises were formed as a result of combat experience in Vietnam. Instead of USAF fighter pilots training in mock combat with pilots flying the same aircraft as their own and using the same tactics, they now face the Aggressors. Flying the Northrop F-5E with a similar performance to a late-model MiG-21, Aggressor pilots are trained in Soviet-style tactics and they operate under strict ground control, as would Soviet pilots. This Dissimilar Air Combat Training prepares US pilots for combat with Soviet fighters having very different characteristics from their own aircraft. Aggressor squadrons are assigned to the USAF in Europe and the Pacific, so that all front-line fighter units can benefit from this training. A further element in USAF combat training is supplied by the Air Combat Maneuvering Instrumentation (ACMI) range. This is a system for recording air combats as they happen, which enables the engagement to be played back during debriefing sessions.

The effective employment of air power will finally depend on the soundness of the tactical doctrines developed by the Warsaw Pact and NATO air forces. Both powers emphasise the role of tactical air forces in support of ground forces and navies. The USAF has tended to place greater reliance on the strategic bomber than the Soviet Union, which favors the ICBM and rather neglects its manned bomber force. This being so it is hardly surprising that the Soviet Union places the greatest importance on air defense, while the USAF has tended to de-emphasize this role. The United States is a pre-eminent exponent of naval air power, a position which the Soviet Union apparently wishes to challenge. However it has much leeway to make up in developing a mature naval air arm based on multi-role air wings and carrier battle groups.

Air support of ground forces will be a major operational commitment for both NATO and Warsaw Pact air forces. However, tactical doctrines differ as to the best means of applying air power to influence the land battle. Both sides see the value of interdiction. Yet whereas NATO regards the aircraft as a quick and potent means of delivering fire support for troops in contact with the enemy, the Warsaw Pact prefers to make use of artillery for this purpose. Air superiority is an important mission for the air forces of both alliances, but again the Warsaw Pact places a good deal of trust in ground based air defenses. The USAF has considerably refined its operational procedures as a result of extensive combat experience in Vietnam and many of the lessons learned over the jungles of Southeast Asia have been applied to the European theater. Apart from its specialized counter-insurgency operations in Afghanistan, the Soviet air force lacks recent combat experience.

Of course the question asked at the beginning of this chapter is unanswerable. Only the acid test of actual combat will show whether the technologically-advanced and individualistic air forces of the West can succeed in mastering the monolithic air armadas of the Warsaw Pact. What is certain is that the outcome of the air battles of World War 3 – influencing and in turn being materially affected by the fighting on land and at sea – will be of decisive importance to the future of the civilized world.

INDEX

Page numbers in italics refer to illustrations

Air defense 33, 56, 64, 66, 117-21, 125-9
  ground environment (ADGE) 117, 121, 127, 129
  maritime 94, 96, 102, 108
  Soviet 47, 190
air refueling 10, 16, 25, 26, 27, 28, 36, 56, 94, 108, 129, 151, 161-2, 184, 186
air superiority 56, 59, 66, 70, 72-5, 78-9, 81, 108, 120, 189
airborne command posts 24-6, 36, 100, 109, 161
airborne early warning 94, 96, 100, 112, 116, 117, 128-9, 129
airborne warning and control system (AWACS) 56, 94, 112, 117, 129
AIRCRAFT see also helicopters
  AMERICAN
  A-3 Skywarrior 147
  A-4 Skyhawk 83, 96, 108
  A-6 Intruder 61, 85, 96, 98, 99, 101, 108, 112, 113
  A-7 Corsair 56, 66, 80, 88-9, 96, 97, 99, 108, 112, 113
  A-10A Thunderbolt 55, 56, 57, 71, 80, 88, 91, 179, 181
  A-37 trainer 88, 90
  AV-8A Harrier 89, 106, 108, AV-8B 89-90
    see also Harrier under aircraft, British
  Advanced Technology Bomber 22-3
  B-1 17, 19, 20, 22, 27
    B-1A 6, 15, 16, 19, 33
    B-1B 10, 19, 22-3, 33, 34
  B-29 Superfortress 30
  B-52 Stratofortress 8-11, 10, 13-14, 15, 16, 19, 23, 26, 27, 30, 33, 34, 79, 82, 112, 185
    B-52D 8, 11, 14-15, 113
    B-52G 8, 9, 9, 10, 10, 14, 33
    B-52H 8, 9, 10, 12, 14, 15
  B-58 Hustler 15
  B-70 Valkyrie 16, 19
  Boeing 707 129, 161
  Boeing 747 43
  C-5A Galaxy 38, 82, 113, 150, 150, 151-2, 153, 153, 158, 159
  C-7 Caribou 158, 158, 168
  C-9A Nightingale 158
  C-17 transport 153, 156
  C-123 Provider 155
  C-130 Hercules 82, 89, 113, 151, 153, 155, 155-6, 158
    AC-130 gunship 88, 89
    DC-130 141
    EC-130 36-7, 85
    KC-130 108
    MC-130 159, 160
    NC-130 44
  C-135 tanker 118
    EC-135 FCP 10, 24, 36
    KC-135A tanker 10, 12, 14, 16, 36, 132, 145, 161-2, 163
    NKC-135A 53
    RC-135 ELINT 25, 132, 142-3, 145, 147
  C-141 Starlifter 150-51, 151, 152-3, 154, 158, 163
  CSIRS fighter 61
  E-2C Hawkeye 96, 100, 100, 112
  E-3A Sentry 112, 117, 127, 129
  E-4 airborne command post 25, 26, 36

  EA-3B Skywarrior 101, 147
  EA-6B Prowler 85, 96, 101, 108
  F-4 Phantom II 56, 59, 61, 65, 72, 74, 78, 79, 82, 89, 99, 117, 120, 120, 121, 182
    F-4B 70, 97
    F-4C 70, 79, 121, 137
    F-4D 56, 70, 73, 79
    F-4E 57, 61, 70, 72, 78, 78, 79, 84, 117, 121, 125, 141
    F-4F 73
    F-4G 57, 70, 84, 137
    F-4J 96, 97, 108
    F-4N 96, 97, 108
    F-4S 96, 97, 108
    FG Mk 1, FGR Mk 2 121
    RF-4B 96, 101, 108, 137
    RF-4C 70, 73, 133, 137, 141, 144
    RF-4E 141
  F-5 fighter 70, 75, 189, 190
  F-14 Tomcat 79, 87, 95, 96, 96, 99, 101, 125
    F-14D Supertomcat 97
  F-15 Eagle 52, 56, 56, 57, 70, 70, 71, 72, 73, 75, 78, 82, 87, 116, 117, 133, 185, 186
    F-15A 56, 57
    F-15C 57, 186
    F-15E 61, 81
  F-16 Fighting Falcon 19, 56, 56, 60, 61, 70, 72, 74, 75, 78, 82, 87, 89, 137, 185, 186, 188
    F-16A 57, 61, 71, 186
    F-16B 74
    F-16XL 185
  F-18 Hornet 79, 81, 95, 97, 108, 120
  F-100F 84
  F-101 Voodoo 116, 117, 120, 121, 144
  F-104 Starfighter 59, 64, 75, 75, 188
  F-105 Thunderchief 84
  F-106A Delta Dart 116, 117, 120, 121
  F-111 strike fighter 15, 16, 41, 56, 59, 60-61, 63, 66, 81
    F-111A 57, 84
    F-111D 56, 63
    F-111E 57
    F-111K 61
    EF-111A 84-5, 186
  FB-111 bomber 13, 13, 14-15, 15-16, 60
  KA-6D tanker 96, 99
  KC-10 Extender 27, 161, 162
  O-2 FAC 89
  OA-37, OA-4M FAC 89
  OV-1 Mohawk 114, 176, 177
  OV-10A Bronco 89
  P-3 Orion 100, 109, 111, 112, 113, 147
  P-51 Mustang 89
  Piper Enforcer 89
  RA-5C Vigilante 101
  RF-8G Crusader 100
  RU-21 Ute 176
  S-3A Viking 96, 98, 99-100, 113
  SR-71 Blackbird 132-3, 134-5, 162
  T-37 trainer 90
  TR-1A stand-off recce 137, 186
  U-2 recce 44, 132, 132, 133, 136
  ANGLO-FRENCH
  Jaguar 41, 59, 63, 81, 82, 90, 130, 141
  ANGLO-GERMAN-ITALIAN
  Tornado 29, 41, 59, 64, 66, 81, 82, 113, 188

  F Mk 2 64, 118, 125
  GR Mk 1 61, 64, 65
  BRITISH
  Buccaneer 61, 73, 112
  Canberra 138
  Harrier 82, 83, 89-90, 90, 101-2, 106, 108
    Sea Harrier 102, 102, 104
  Hawk trainer 90, 127
  Jaguar see aircraft, Anglo-French
  Lightning 116, 121
  Nimrod 110, 112, 129, 129, 147
  Shackleton 129
  Tornado see aircraft, Anglo-German-Italian
  VC10 148, 156, 157
  Victor 163
  Vulcan 21, 29, 61
  CANADIAN
  CF-101 Voodoo 116, 120
  CHINESE
  A-5 Fantan 189
  B-6 bomber 30
  Tu-4 nuclear bomber 30
  FRANCO—GERMAN
  C.160 Transall 156, 157
  Alpha Jet 90, 122, 181
  FRENCH
  C-135 F tanker 29, 30
  Alizé 101, 101
  Atlantic 112, 113, 147
  Mirage III 30, 75, 76, 102, 121, 138, 141
  Mirage IV 21, 29, 30
  Mirage 5 75
  Mirage 2000 75, 78, 115, 121
  Mirage 4000 31, 115
  Mirage F1 118, 121, 141
  Noratlas 156
  Super Etendard 93, 101, 112
  GERMAN
  Alpha Jet 90, 122, 181
  Fiesler Fi 103 (V1) 33
  ITALIAN
  Macchi MB326 90
  JAPANESE
  Shin Meiwa PS-1 112
  SOVIET
  Antonov An-12 85, 147, 154, 184
    Cub 85, 147, 154, 156, 157
  Beriev Be-12 112
  Ilyushin Il-18 112, 147
    Il-38 May 112
    Il-76 Candid 27, 154, 154, 184
    Il-86 airliner 27, 129
  Mikoyah-Gurevich MiG-17 189
    MiG-19 189
    MiG-21 8, 58, 70, 72, 78, 79, 81, 137, 141, 185, 186, 189
    MiG-23/27 Flogger series 58, 185, 186
    MiG-23 59, 78, 78, 79, 81, 91, 121, 125, 186, 189
    MiG-27 59, 66, 79, 91
    MiG-25 Foxbat 16, 121, 125, 133, 141, 185
    MiG-29 79
  Mil Mi 14 112
  Myasishchev M-4 Bison 18, 25, 26, 27,
    M-50 bomber 27
  Ram aircraft 45, 79, 91, 136
  Sukhoi Su-7 Fitter 58, 91
    Su-9 121
    Su-11 Fishpot 121
    Sub-15 Flagon 121, 125
    Su-17 Fitter 58, 59, 91, 91, 109, 113
    Su-22 91
    Su-24 Fencer 29, 58, 59, 66, 66, 81

  Tupolev Tu-4 30
    Tu-16 Badger 18, 25, 27, 28, 28-9, 40, 41, 113, 147, 182
    Tu-20 Bear 17, 23, 25, 26-7, 33, 112, 146-7, 147
    Tu-22 Blinder 18, 25, 28, 33, 41, 113
    Tu-26 Backfire 22, 25, 27-8, 29, 33, 41, 66, 113, 183, 187
    Tu-28 Fiddler 121, 125, 129
    Tu-95 see Tu-20 Bear
    Tu-114 airliner 129
    Tu-126 Moss 128, 129
    Tu-144 airliner 27
  Yakovlev Yak-25 Mandrake 136
    Yak-28 Brewer-E 85, 85
    Yak-28 Firebar 121, 125
    Yak-36 Forger 104, 106, 113
  SWEDISH
  Lansen fighter 138
  Saab 105 trainer 90
aircraft, carrier-borne 61, 70, 85, 88, 93, 94, 96-7, 99-102, 104-6, 147, 184
aircraft carriers 8, 94, 96-7, 99-102, 104, 113, 185
airfield defense 82-4, 125
anti-aircraft artillery 14, 56, 59, 82, 83, 84, 87, 178, 186
anti-ballistic missile systems 44, 53, 116
anti-satellite weapons 44, 52, 53
anti-submarine warfare 94, 96, 99-100, 104, 105, 106, 109-12
anti-tank role 168, 169, 172, 176, 178, 181
attack fighters 91, 94, 96-7, 99, 112

Ballistic missile defense 38, 121
  early warning 44, 44, 127-9
Belgian air force 72, 75, 87, 126, 187
bomber aircraft 8-16, 18, 22-3, 25-7, 29, 30, 33, 161
bombs 66-7, 81-2, 84, 87, 88, 112
British Army 174, 181

Canadian air forces 75, 120, 121, 174
China 47, 121, 189
  nuclear weapons of 31, 41
close air support 56, 66, 75, 79, 87-91, 97, 99, 108, 168
command, control and communications 36, 37, 116, 117, 128-9, 161
counterair see air superiority

Danish air force 72, 75, 126

Electronic countermeasures 9, 14, 16, 19, 22, 33, 56, 61, 64, 66, 74, 84-5, 101, 113, 137, 145
electronic intelligence 44, 47-8, 100, 101, 132-3, 144-5, 147, 176-7
electronic warfare 84-5, 94, 96, 101, 108, 168, 186

Falkland Islands 101, 102, 105, 112
foward air control 56, 89
French air force 29-30, 75, 90, 112, 121, 126, 141, 156, 169, 181, 187
French navy 30, 101, 106

German air force 61, 64, 66, 67, 70, 75, 82, 90, 112, 121, 126, 141, 156, 174, 176, 181, 187
German army 169
German navy 64, 113, 147
glide bombs 64, 67
Greek air force 70, 75, 126, 174
ground attack see close air support

Head-up display (HUD) 69, *69*, 72, 74, 89, 87, 108
helicopters 105, 159, 168-9, 172-81, 185
 AMERICAN
 Ah-1 106, 109, 177-8, *178*, 179, 180, 181, 185
 AH-64 Apache 167, 176, 179-80, *179-80*, 181
 CH-3 141, 161
 CH-46 Sea Knight 109
 CH-47 Chinook 173, 174, 177
 CH-53 106, 113, *113*, 161, 174, *175*
 CH-54 Tarhe 173-4, *174*
 EH-60 177
 HA-53 161, *164*
 MH-53 Super Stallion 113
 OH-6A Cayuse *175*, 176
 OH-58 Kiowa 176, 179
 RH-53 161
 SH-2F Sea Sprite 105, *105*, *108*
 SH-3 Sea King 96, 99, 100, 101, 102, *106*
 SH-60B Sea Hawk 105
 U-60A Black Hawk 105, 168-9, 172, 177
 UH-1 Huey 106, 109, 161, 168, 169, *169*, 172, 173, 177
 ANGLO-FRENCH
 Gazelle 109, 169, 176, 181
 Puma 169, 174
 BRITISH
 Lynx *104*, 105-6, 109, 181
 Westland Scout 180
 Westland Sea King 109
 Westland Wessex 109
 FRENCH
 Alouette 101, 169, 181
 GERMAN
 Bö 105, 176, 181
 SOVIET
 Helix 106, 108
 Ka-25 Hormone 104, 106, 108
 Mi-1, Mi-2, Mi-4 176
 Mi-6 Hook *169*, 172, 175-6
 Mi-8 Hip 169, 172, 173, *173*, 181
 Mi-10 Harke 175
 Mi-24 Hind 109, 172, *172*, 181
 Mi-26 Halo 176
helicopters, roles of 88, 108-9, 172-3, 176, 177, 178, 181, 186

Identification, Friend or Foe (IFF) system 96, 125
Inertial navigation 16, 64, 69, 89, 99, 113, 137, 141
infra-red 52
 counter measures 178, 180
 forward-looking (FLIR) 9, 61, 69, 88, 97, *97*, 99, 108, 111, 178, 179, 180, 186
 imaging equipment 68, 176
 linescan sensor 137, 141
 missile guidance 16, 67, 79, 81, 87, 121, 125
interceptors 8, 9, 14, 26, 33, 75, 79, 116-17, 120-21, 125, 127, 185
interdictor/strike 29, 56, 58-61, 66, 79, 108, 186
Italian air force 61, 64, 75, 112, 113, 126, 169, 174

Laser
 imaging equipment 67, 68
 laser weapons 53
 range finder 64, 69, 79, 91, 176, 181
 target designation 69, 179, 186
 weapon guidance 67, 68, 69, 91, 176

Magnetic anomaly detector (MAD) 99, 100, 102, 105, *105*, 106, 111, 112

marine surveillance 14, 27, 48, 112, 129
medical evacuation 153, 158, 168, 172
minelaying 14, 30, 99, 113
missiles, air-to-air 9, 70, 75, 79, 125, 186
 AA-2 Atoll 66, 78, 79, *79*, 81, 91
 AA-3 Anab 125
 AA-6 Acrid 125
 AA-7 Apex 78, 79, 81
 AA-8 Aphid 66, 78, *78*, 79, 81
 AIM-4 Falcon 121
 AIM-Sparrow 70, 72, 79, *79*, 81, 84, 96, 97, 125
 AIM-9 Sidewinder 61, 66, 70, 72, 75, 79, *79*, 81, 84, 89, 96, 97, 99, 102, 109, 121, 181
 AIM-54 Phoenix 79, 82, *95*, 96, 97, 125
 Advanced medium range 74
 Matra R530 76, 121
missiles, air-to-surface 26, 33, 67, 70, 81, 82, 112
 AGM-45 Shrike 84
 AGM-65 Maverick 61, 67, 74, *80*, 84, 88
 AGM-78 Standard 84
 AGM-88 HARM anti-radiation 84
 AM-39 Exocet 101, 112
 ASMP tactical nuclear 101
 Soviet AS series 28, 66, 68, 78
missiles, anti-ballistic missile 121, *124*, 127
missiles, anti-radiation 30, 66, 67, 84, 99, 186
missiles, anti-ship 48, 112, 113
 AGM-12 bullpup 112
 AGM-84 Harpoon 105, 112
 Kormoran 113
 SS-N-12 106
 Sea Eagle 102, 113
 Sea Skua 105
missiles, anti-tank 66, 67-8, 109, 169, 172
 AT-2 Swatter 181
 BGM-71 TOW 178, 179, 181, 185
 Hellfire 176, 179
 HOT 176, 181
missiles, cruise 8, 9, 22, 27, 28, 30, 33, 34-5, 48, 127, 185
 air-launched 35-6, 37, 183
 AGM-86 8, *23*, 27, 33, 34, 82
 ground-launched 34, 41
 AGM-109 Tomahawk 34, 82
 P4T (British) 82
missiles, intercontinental ballistic (ICBM) 8, 25, 30, 35, 37, 52, 145
 Atlas 37
 Chinese missiles 41
 MX 32, 36, 37-8, 127, 184
 Minuteman *26*, *30*, *34*, 36, 37-8, 184
 Polaris 29
 Soviet SS series *38*, *39*-41, 183
 Titan *31*, *35*, 37
 Trident 29
missiles, intermediate range 30, 41
 Pershing *40*, 41
 SS-4 *39*
missiles, short-range attack 8, *13*, *14*, 16, 19, 33
missiles, submarine-launched 8, 23, 30, 37, 110, 128, 183, 184
missiles, surface-to-air (SAM) 9, 14, 33, 56, 59, 82, 83, 84, 87, 96, 121, 125, 185, 186
 Bloodhound 126
 Hawk *122*, 125-6
 Nike Hercules 125
 Rapier *122*, 126
 Soviet SA series 84, 121, *124*, 126-7, 178
 Stinger 181

NATO forces 41, 56, 57, 58-9, 61,

68, 70, 72, 75, 89, 90, 105, 108, 111, 112, 120, 126, 128-9, 141, 155-6, 169, 174, 176, 184, 186-9
navigational aids 9, 16, 33, 64, 68-9, 102, 133, 137
Netherlands air force 72, 75, 112, 126
North American Aerospace Command (NORAD) 117, 120, 127
Norwegian air force 75
nuclear weapons delivery 8, 23, 25, 27, 28, 29, 30, 33, 59, 66, 97, 99, 101, 110, 127, 137, 183, 184

Photographic reconnaissance 94, 96, 132-3, 137, 141, 144

Radar 9, 19, 22, 68, 74, 81, 96-7, 99, 101, 110, 116, 117, 145, 177, 186
 altimeter 9, 16, 68, 137
 counter measures 85, 87, 101, 125, 177, 186
 doppler 16, 22, 64, 72, 79, 89, 91, 96, 133, 137
 forward-looking 9, 133, 137
 missile guidance 67, 79, 81, 84, 96, 113, 121, 125, 126, 186
 radar signature 9, 22
 side-looking 44, 101, 133, 137, 141, 144, 147, 176
 target acquisition 16, 81, 125
 terrain avoidance 9, 68, 91, 99
 terrain following 9, 16, 19, 64, 66, 69, 88
reconnaissance 25, 28, 38, 44-5, 48, 52, 56, 75, 108, 132-3, 136-7, 141, 144-5, 147, 161
remotely piloted vehicle (RPV) 84, 87-8, 141
rockets, unguided 67, 112, 121, 177, *180*
Royal Air Force 29, 58, 61, 64, 66, 67, 70, 84, 89, 90, 110, 112, 121, 126, 129, 141, 147, 151, 156, 168, 174, 187
Royal Navy 83, 101-2, 105, 109

Satellites
 American 44, 47-8, 50-52
 ballistic missile warning 44, 50, 52
 communications 11, 36, 44, *46*, 48, 50, *50*
 ELINT gathering 44, 47-8
 meteorological 44, *47*, 48
 navigation 44, 50, 69
 reconnaissance 38, 44-5, 132, 147
 Soviet 44, 45, 48, 50, 52
 surveillance 44, 48, 112, 128
sonar, sonobuoys 99, 100, 101, 102, 105, 106, *108*, 110-11, 112
Soviet forces *passim*
 Frontal Aviation 23, 25, 29, 57-8, 66, 68, 78-9, 85, 90, 91, 136, 161, 184, 186
 Long Range Aviation 23, 25, 27, 29, 30
 Naval Air Force 25, 29, 109, 113, 147
 Navy 48
 Rocket Forces 23, 37, 47
 Transport 25, 153-4, 156, 159, 161
space shuttle *43*, 44, 47, 52-3
Spanish air force 70, 75, 121, 126, 169, 174
'Stealth' technique 10, 22-3, 34, 61, 84
STOL 82, 83, 89, 101, 104, 105, 108
Swedish air force 83

Tanker aircraft 10, 11, *12*, 16, 27, 56, 99, 184

training 13, 75, 90, *96*, 190
transport aircraft 82, 108, 150-56, 158-9, 161-2, 184
Turkish air force 70, 75, 169, 187

United States air forces *passim*
 Aerospace Defense Command 117
 Air Defense Command 26, 117
 Air Force Reserve 13, 56, 57, 151, 155, 161, 187
 Air National Guard 13, 26, 56, 57, 84, 88, 117, 120, 155, 156, 161, 187
 Military Air Lift Command 150, 155, 184
 PACAF 56, 57, 137
 Space Command 44
 Strategic Air Command 8, 10-16, 33, 36-8, 48, 50, 117, 132, 161
 Tactical Air Command 56-7, 72, 117, 155, 184, 190
 USAFE 56, 57, 121, 187
United States, other forces
 Army 168, 172, 173, 176, 177, 178, 180
 Civil Reserve Air Fleet 151, 184
 Marine Corps 83, 89, 97, 101, 108, 109, 137, 184, 187
 Navy 8, 34, 36, 48, 50, 60, 61, 70, 85, 88, 94, 96-7, 99-102, 105, 108, 113, 147, 184, 185, 187
Rapid Deployment Force 14, 150

VTOL 89, 104-5, 108

Warsaw Pact allies 56, 58, 59, 121, 183, 186, 188-9, 190

The publishers would like to thank Adrian Hodgkins who designed this book, and R. Watson who compiled the index. The Public Affairs and Audiovisual Departments of the US Department of Defense, the US Army the US Air Force, the US Navy and the US Marines supplied many of the pictures and in addition thanks are due to the following for illustrations.

Avions Marcel Dassault: pp 20 bottom, 76, 92-93, 100 bottom, 113 left, 114-115, 119, 123, 138 top.
Bison Picture Library: pp 38, 39, 48, 49 bottom, 59, 77, 85, 137.
British Aerospace: pp 4-5, 20-21, 65 top, 86, 90, 103 top, 104 bottom, 106-107 top, 110 top, 118-119, 120 center, 122-123, 128 bottom, 129, 130-131, 136 both, 138-139, 148-149, 162-163.
Michael Badrocke: artwork pp 20 top, 102-103, 152-153.
Canadair: p 140 all three.
Peter Ensdleigh Castle: p 28-29 (artwork).
General Dynamics: pp 60-61, 74 both, 185.
German Navy: p 75.
Grumman: p 94-95.
Hughes: pp 166-167, 179.
Lockheed: p 163.
Martin Corporation: pp 31, 32, 40-41, 49 top, 51.
Messerschmitt-Bolkow-Blohm: pp 68-69, 171 bottom.
Ministry of Defence (London): pp 116 bottom, 146 bottom.
Rockwell International: pp 17 top, 64.
SAAB Scania: p 139 top.
VFW-Fokker: pp 157 top, 174-175.
Vought Corporation: p 66-67
Westland Helicopters: 170 bottom.